U, THE UNIVERSE AND ETERNITY

How Science and Religion are Interrelated

ALL THE BEST
1958

U, THE UNIVERSE AND ETERNITY

How Science and Religion are Interrelated

Stephen R Sopher

ATHENA PRESS
LONDON

U, the Universe and Eternity
How Science and Religion are Interrelated
Copyright © Stephen R Sopher 2007

All Rights Reserved

No part of this book may be reproduced in any form
by photocopying or by any electronic or mechanical means,
including information storage or retrieval systems,
without permission in writing from both the copyright
owner and the publisher of this book.

ISBN 10-digit: 1 84748 004 7
ISBN 13-digit: 978 1 84748 004 0

First Published 2007 by
ATHENA PRESS
Queen's House, 2 Holly Road
Twickenham TW1 4EG
United Kingdom

Every effort has been made to trace the copyright holders of works quoted within this book to obtain permission. The publisher apologises for any omissions and is happy to make necessary changes in subsequent print-runs.

Printed for Athena Press

Preface

This book attempts to consolidate and to convey the significance of the extraordinary advances in the various fields of science including geology, physics, chemistry, biology and cosmology, and highlight their impact on our thought processes.

We are beginning to come up with the answers to the age-old questions of when and how things got started in the universe, eventually leading to the origin of life on our planet.

We are also reminded of how our cultural traditions have influenced and continue to influence our thought processes, and how religion sought to provide us with answers.

It is not my intention to offend any particular belief or faith in this study, but to provide the basis for conciliation in the global community.

My background in geology and physics provided the necessary impetus for pursuing the subject matter. My appreciation of cultural influences came from the circumstance of being born and raised in China in combination with having lived or visited all the continents, except for Antarctica.

Finally, I would like to dedicate this book to my father and uncle, both now deceased, who influenced me from a very early age and on to adulthood by their philosophical and scientific knowledge and their approach to life in general. At the same time, I would like to acknowledge my appreciation for the encouragement given by Marianne, my wife.

Contents

List of Illustrations

Introduction

This work has been inspired by the giant steps that have been taken in the twentieth century by workers in the diverse fields of science. This trend is expected to accelerate in the twenty-first century, leading to a more comprehensive understanding of the fundamental concepts of physics and biology with reference to the mysteries of space and time and our modest place in the universe.

Gazing across the universe over an expanse of billions of years with the most sophisticated instruments has already given us a better understanding of the origin and structure of the universe. Nevertheless, many key questions such as those regarding the composition of so-called 'dark matter', as well as the composition of the highest energy cosmic rays, require continued integration of theory with observational facts.

Observing and understanding the most probable physical processes involved in the continuing development of our planet Earth has been extremely rewarding. In the field of microbiology, the self-assembly of the first cells capable of growth and division leading to the initial appearance of the least complex forms of life has been postulated. The sequence of events leading to the eventual appearance of more complex creatures we see today, including the human species, is captured in the fossil record.

As human thought has developed, reaching out to contemplation of the universe, mankind's geocentric position has shifted and humanity has come to realise and appreciate its modest status within the universe.

In the development of religious concepts, the gradual transformation from idol worship to the revolutionary idea of monotheism has shaped the culture of Jews, Christians and Muslims alike. The recognition of a God or transcendent being has become a matter of faith; however, it is suggested that the essence of this ultimate spiritual concept may lie in its unity with nature and nature's fundamental forces.

It is envisaged that through a conscientious partnership of science, theology and philosophy, our ethereal origins, our present state and our eternal future can be illuminated.

I

Beginnings

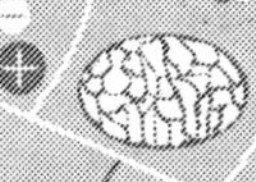

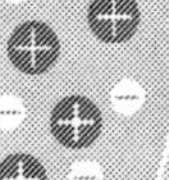

Big Bang
Time
10-35s
10-32s
10-10s
300s
300,000 years
1 billion years
15 billion years
Superstring (?) era
GUT era
Inflation era
Electro-Weak era
Particle era
Recombination era
Galaxy and star formation
Present era
S
E
W
A period when all force and energy was indistinguishable, still speculative
Particles responsible for strong, weak, and electromagnetic forces emerge
Universe grows from atomic to cosmic scale in a thousandth of a second
Electromagnetic and weak forces become independent
Protons and electrons form
Protons join with electrons; the first light from Big Bang emerges

Wonder and Mystery

In the eternal journey through space and time, our earthly experience is filled with wonder and mystery. The essence of this mystery is brilliantly captured in the following lines by Carruth:*

> A FIRE-MIST and a planet,
> A crystal and a cell,
> A jellyfish and a saurian,
> And caves where the cavemen dwell;
> Then a sense of law and beauty
> And a face turned from the clod –
> Some call it Evolution,
> And others call it God.

Big Bang

It is contended that religion gains by embracing the fields of intellectual knowledge provided by science and that through a possible joint endeavour both religion and science could reconcile themselves by acquiring humility and retaining their respective states of self-confidence.

The joint search for understanding the depths of the subconscious with all its ramifications is of fundamental importance. A reconciliation between rational truth and what may be called intuitive truth and, in turn, their ultimate relation to an underlying fundamental truth will be necessary.

From recent developments in science, we are only now beginning to comprehend that our universe started with the big bang, estimated to have taken place between 13 and 14 billion years ago. Our universe is perceived to have been initially composed of

---

* Carruth, 'Each to His Own Tongue', *The New England Magazine Co.*, vol. 19, 1895, p.323.

hydrogen and helium, the smallest atoms. The other elements are believed to be the products of nuclear reactions occurring in stars and supernova explosions. The solar system was eventually formed some 4.6 billion years ago, and the age of our planet, Earth, is estimated at 4 billion years.

At this early juncture, it is cautioned that one should not be complacent about these points. With the accelerating pace of research and discovery, new interpretation and/or reinterpretation of evidence will surely occur. Mirages resulting not only from refractive images, but from points of view will have to be addressed and resolved. Potential parallel universes or 'multiverses', expanding away from ours at super-speeds and corresponding to pre- or post-big bang, may very well be in the realm of possibility.

Road Map of the Universe

In the road map for the Structure and Evolution of the Universe (SEU) prepared by NASA, three fundamental, scientific quests are listed. They are:

- to explain structure in the universe and forecast our cosmic destiny;
- to explore the cycles of matter and energy in the evolving universe; and
- to examine the ultimate limits of gravity and energy in the universe.

These quests will be developed by focusing on the following research programs, which will:

- identify dark matter and learn how it shapes galaxies and systems of galaxies;
- explore where and when the chemical elements were made;
- understand the cycles in which matter, energy, and mag-

netic field are exchanged between stars and the gas that lies between stars;

- discover how gas flows in discs and how cosmic jets are formed;
- identify the sources of gamma-ray bursts and high-energy cosmic rays; and
- measure how strong gravity operates near black holes and how it affects the early universe.

Initial Observations

This ambitious road map, prepared in 1999, is already bearing fruit.

With the launch of the Chandra X-ray telescope, NASA's X-ray eye in the sky, astronomers are receiving cosmic images that will give them a better perspective on ancient astral events, including the explosion 1,000 years ago of a star about ten times larger than the sun. It appears that super-massive black holes once dominated the universe, sucking in gas, dust and whole stars, and erupting with surges of X-rays that have journeyed since for billions of years across the universe.

One study discovered that an exploding star 200,000 light years from Earth is spitting huge amounts of oxygen into space. That oxygen might serve as the raw material to fuel the creation of new stars and planets.

In other observations, researchers peered into the heart of the Milky Way galaxy, where a giant black hole is thought to consume hot gas and stars. It was found, however, that the region around the black hole does not glow as brightly as scientists had thought it would. It was thought that the matter spiralling into a black hole would be heated to tens of millions of degrees, causing it to virtually blaze in X-rays. This was not found to be the case, so it is inferred that something unusual is occurring there, which may be blocking the X-rays.

Using the Hubble Space Telescope as well as the Keck I Telescope in Hawaii, an international team of astronomers studied an ancient star in the halo of the Milky Way some 2,500 light years from Earth. From the radioactive decay of elements such as lead, the team estimates the star is almost 15 billion years old – three times as old as the sun. Other elements such as gold and other precious metals were also identified.

At Johns Hopkins University, scientists have indicated that the initial colour of the universe was green to turquoise and that the ultimate colour, occurring in a few billion years, will be red.

This recalls the beauty, tranquillity and harmony of the rainbow here on Earth. This phenomenon is caused by the internal reflection and refraction of the sun's rays in the spherical globules of raindrops. The colours of the rainbow are those of the spectrum in the same order, with the violet rays (the most refracted) on the inside and red on the outside. With observation from a sufficiently high altitude above the raindrops, and with the sun at the right angle, the rainbow appears as a concentric circular structure.

The rainbow was well known in ancient times. In the conclusion of the Flood narrative in Genesis, it is acknowledged as a sign of the covenant between God and Noah and all living creatures. In Hebrew, it is the word for the weapon, the *bow*. The God of Israel, as well as other ancient Near Eastern deities, was depicted in the Old Testament as a warrior god, especially in his role as god of the storm. Lightning bolts are his arrows, which he shoots from his bow. After the Flood, the rainbow became a sign that the deity of the storm will never again use this weapon for total destruction.

Other investigations in cosmology include the study of the geometry of the universe. Observations from the Boomerang Telescope over Antarctica and the Maxima Telescope over Texas appear to indicate a 'flat' universe. Not flat in the conventional sense; flatness here means that space as a whole is not curved.

Other data suggest the balance of mass and energy in the universe is as follows: 5% is normal matter that makes up galaxies, stars, planets, interstellar dust, quasars and other astronomical objects; 30% is 'dark matter', the identity of which is one of the great remaining mysteries of cosmology; and 65% is the equally mysterious 'dark energy'.

Prior Investigations and Theories

When Albert Einstein developed the general theory of relativity in 1915, he believed the universe was unchanging. To make his calculations fit this concept of an unchanging universe he introduced a 'cosmological constant'. In 1917, the Dutch astronomer Willem de Sitter (1872–1934) excluded this constant to allow for the possibility that the universe might be constantly expanding.

It was not until the 1920s that astronomer Edwin Powell Hubble discovered that the spiral-shaped objects astronomers had seen in the sky were, in fact, other galaxies.

The Belgian astronomer and Jesuit priest, Georges Henri Lemaître, is credited with being the first to suggest that the universe came into being with a big explosion some 15–20 billion years ago. (It is to be noted that the Vatican Observatory at Castel Gandolfo was built in the sixteenth century on the authority of Pope Gregory XIII to help Jesuit astronomers assess the need for calendar reform. From charting the sun's movement along the walls and across a meridian line marked on the floor, the Jesuit astronomers were able to calculate how far the Julian calendar had deviated. This was corrected by simply erasing ten days from the old calendar in 1582.)

In 1929, Hubble found observable proof that all matter in space is moving apart and that some galaxies were moving at enormous speeds. This gave the necessary proof that the universe was expanding and made the big bang theory more credible.

However, it was cosmologist Alan Guth of MIT who came up with his 'inflation theory' in 1979 to explain what could have caused the rapid expansion of the universe after the big bang occurred. He contends that an early and spontaneous inflation of energy in a so-called 'false vacuum' smaller than a proton not only got the universe started but served to expand it exponentially. Many versions of inflation theory have been conceptualised.

Alternative theories to the big bang are the steady state theory and the plasma theory.

The steady state theory is attributed to astronomer Thomas Gold (1920–) and strongly supported by Fred Hoyle (1915–2001), professor of astronomy at Cambridge University. They accepted the view that the universe was expanding but rejected the premise that it all started with a single cataclysmic event. To Hoyle, the universe was eternal and had always existed, maintaining a constant density and an overall constant appearance.

It is worth noting that Hoyle coined the expression 'big bang' to ridicule his opponents' views of a unique moment of creation. Much to his and his colleagues' chagrin, the name caught on.

From the very beginning, this big bang theory was enthusiastically embraced by the Vatican, as it believed that science had now confirmed the biblical concept of creation and a created universe.

In the plasma theory, it is proposed that the universe was born out of electrical and magnetic phenomena involving plasma. Plasma consists of electrically charged atoms and electrons, at a very high temperature. Plasma cosmologists believe that the evolution of the universe in the past must be explained in terms of the processes occurring in the universe today. They reject the big bang concept and claim that the expansion of the universe is due to the interaction of matter and antimatter. The evolving universe we see today reflects on what happened in the past and will continue to exist and evolve for eternity.

From astronomic observations carried out in the 1920s, Edwin Hubble discovered that certain spiral-shaped objects seen in the sky were in fact other galaxies. In observations by other astronomers it was determined that the galaxies were expanding outward.

Continued observations by Hubble and his assistant, Milton Humason, culminated in a 1931 paper indicating the rate at which the universe was expanding. This rate is now known as Hubble's constant, and calculated to be 558 kilometres per second per mega-parsec. By using this constant, they came up with an age for the universe: 1.8 billion years.

In the 1930s and 1940s, using radiometric dating techniques, geologists arrived at a figure between 3 and 4 billion years for the Earth's age. For the universe to be younger than the Earth was implausible.

It was only after using high-powered telescopes and advanced technology that an age of 13–14 billion years was determined for the universe.

It is generally believed that the universe started as a dense bundle of subatomic particles followed rapidly by the formation of atoms, electromagnetic and other forces, stars and galaxies. Experiments designed to determine the actual make-up of the primordial soup at the moment of creation of our universe are currently being carried out at CERN's Geneva research facility and at RHIC's accelerator on Long Island.

Remarkable images of how our infant universe looked when it was just 380,000 years old were released by NASA in February, 2003. A satellite called the Wilkinson Microwave Anisotropy Probe, which is orbiting the sun 1 million miles father out than the Earth, measures temperature changes down to millionths of a degree. This ability to detect minute changes in temperature made it possible to discern tiny spots and ripples that would eventually grow into the stars and galaxies we see today.

New measurements presented by physicists at Princeton University in March 2006 provided evidence, from variations in the faint cosmic glow of the infant universe, of what transpired in the first trillionth of a second. In that minuscule time after the big bang the universe expanded from a pin point to a volume larger than all of observable space through a process known as inflation.

Evidence indicates that the first stars were formed about 200 million years after the big bang. It was also determined that the early universe had a temperature of 3,000 °C, and as it expanded, it began to cool rapidly. It is conjectured that as the universe expands further at an accelerated rate, it will cool to the point that it will no longer have any energy. According to scientists, this is likely to occur in approximately 13 billion years and, as depicted by Charles Bennett of NASA's Goddard Space Flight Centre in Greenbelt, MD, '...the universe dies in a kind of slow, cold death'.

The Milky Way and our Solar System

Our galaxy, the Milky Way, is one of 50 billion galaxies. The Milky Way, in turn, contains billions of stars, planets, glowing nebulae, gas, dust and empty space. Older stars are generally found near the centre of our galaxy while younger stars lie in the plane of this rotating galaxy.

Around our sun, one of the more modest stars in the Milky Way, our solar system was born. The objects orbiting around the sun include planets, asteroids, comets, meteors and meteorites and other particles.

The planets all orbit the sun in the same direction, which coincides also with the direction of the sun's spin. Of the nine planets, the inner planets (Mercury, Venus, Earth and Mars) are relatively small and solid, while four of the five outer planets (Jupiter, Saturn, Uranus and Neptune) are gaseous giants. Pluto, once considered the farthest and smallest planet, was downgraded by the International Astronomical Union in 2006 and is now categorised as a dwarf planet. Astronomers have speculated about the existence of two other planets: one between Mercury and the sun and Planet X, beyond Pluto.

Information regarding the composition and layering of the interior of our planet, Earth, is obtained using seismic waves which would be either natural vibrations produced by earthquakes or explosions set off artificially. Certain types of seismic waves are deflected when they encounter a change in density. One type of wave, the S-wave, does not travel through liquids. From interpretation of the data, it has been determined that there is a solid inner core with a diameter of 3,000 km composed of iron with some nickel; a liquid outer core of the same composition extending 2,100 km; a so-called 'mantle' composed of dense semi-molten rock 3,000 km thick covering the core; and, finally, a relatively thin silicate crust in the outer shell.

Of all the planets, the Earth has the strongest magnetic field and most active geology. The outer molten core with its dynamo-like motion is the source of the magnetic field. The heat source for the core is derived mainly from the decay of radioactive elements such as uranium and thorium, which are relatively abundant on Earth.

The density of the Earth was first calculated by the English chemist and physicist, Henry Cavendish (1731–1810). Using equipment based on the principle of the torsion balance, he was able to measure the constant force of attraction, the Newtonian constant, G. Using this constant he was able to calculate the density of the Earth at 5.52 g/cm^3. In 1900, with improved and significantly more sensitive instrumentation, the density of the Earth was deemed to be 5.5268 g/cm^3. For additional information on the Earth's parameters, see Appendix C.

However, the density of the rocks on the Earth's surface is only 2.70 g/cm^3. It is therefore thought that there must be a much higher density in the Earth's core, probably composed of iron with some nickel, to explain the overall density of 5.5268 g/cm^3.

Additional evidence from fragments of space meteorites is quite illuminating. There are two main types of meteorites: the stony type, composed of mainly silicate material similar to the Earth's crust; and the iron-rich type, closely resembling what is presumed to be at the Earth's core.

Our closest neighbour, the moon, is located just 250,000 miles from Earth. Its dimensions – about one fourth the diameter of the Earth's – its axis of rotation and orbit around the Earth, have been known for centuries. In addition, its outward appearance, showing an abundance of craters and widespread areas, presumed to be the result of lava flows, is quite evident.

However, it was not until space flights were inaugurated in the 1960s, with astronauts landing on the moon in 1969 and the continuing Apollo missions, that definitive information was obtained. From rock samples collected and geophysical experiments carried out on the lunar surface, a clearer picture of its history and processes is emerging. From samples collected from the lunar highlands, an average age of about 3.9 billion years has been reported. These figures come surprisingly close to the age of the Earth. The craters were found to be the result of meteorite impact. In some cases, these impacts triggered volcanic activity which formed the smooth and extensive 'seas'. Like the Earth, the moon also appears to have a layered structure. The crust averages about 70 km; the mantle extends about 800 km and the core about 400 km. As the moon is reported not to have a detectable magnetic field, the core is believed to be solid at present, although this does not preclude a molten past as evidenced by some magnetised samples collected.

Of the many theories regarding the origin of the moon, the most acceptable at this time is that an asteroid had a massive collision with the Earth some 4.6 billion years ago. The explosion

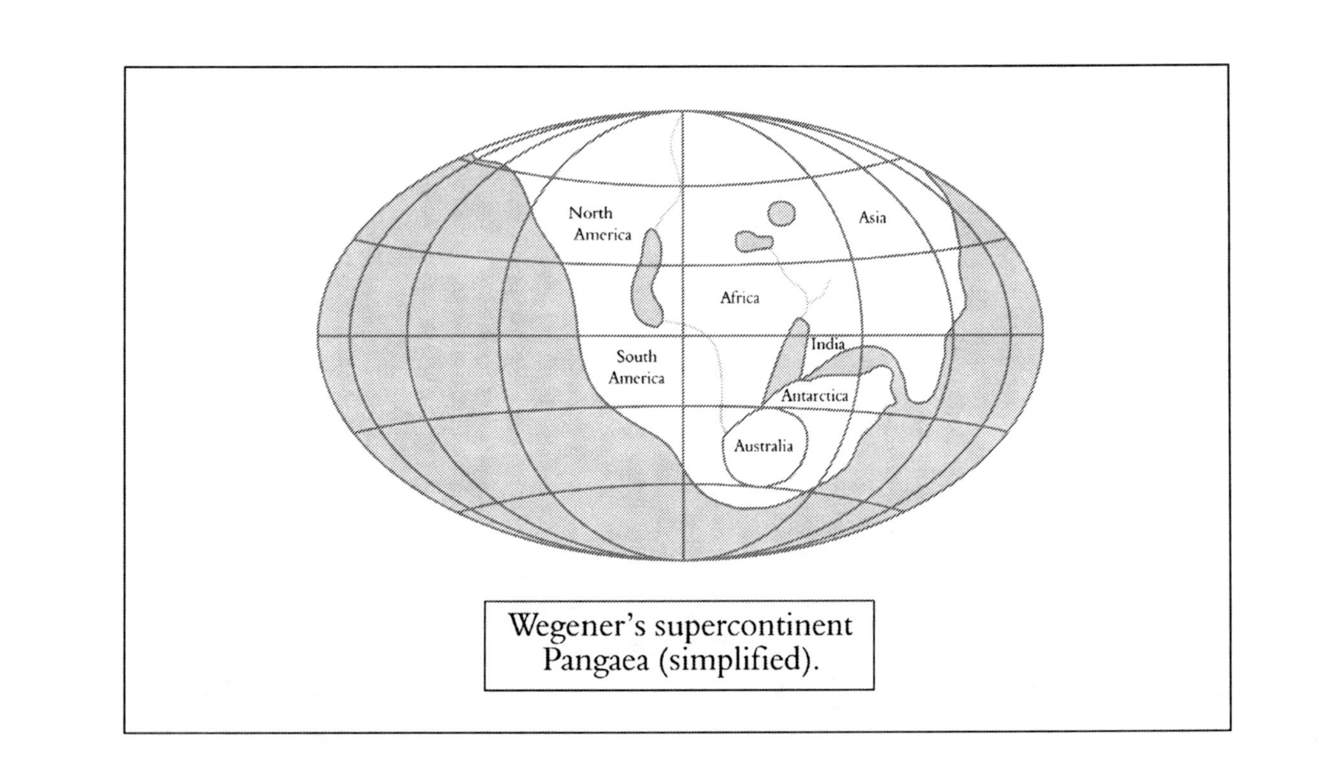

Wegener's supercontinent Pangaea (simplified).

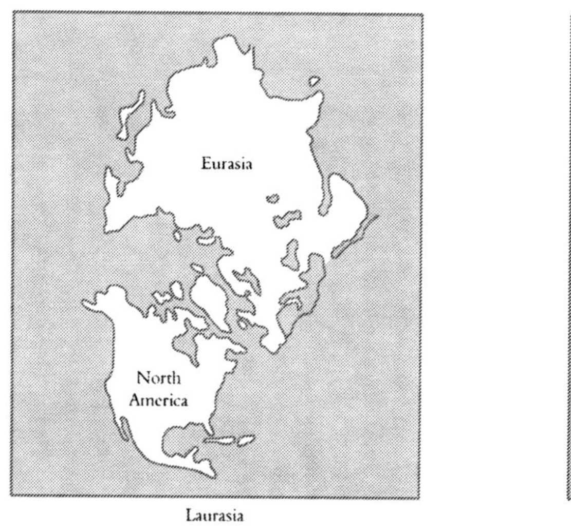

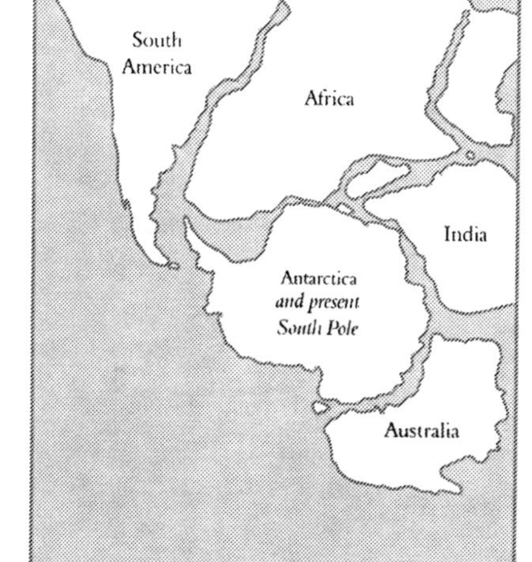

Continental drift - break-up of Pangaea (simplified)

caused a substantial amount of material to be ejected from the Earth. This material included mantle, molten core and other debris. The material coalesced and consolidated into the primitive moon which circled around the Earth.

Our Planet Earth

Coming back and down to Earth and its tumultuous history evidenced by extensive mountain ranges and current volcanic as well as seismic activity, satisfactory explanations are in order.

Until the dawn of the twentieth century, the driving mechanism or mechanisms for the development of continental as well as oceanic phenomena was poorly understood.

It was not until 1910 that a German meteorologist, Alfred Wegener, came up with the hypothesis that the continents have not always been in their present positions. From observing a map of the world, he could visualise that there was a good fit between the east coast of South America and the west coast of Africa. Similarly, India and Asia could be united with the east coast of Africa, and North America could be merged with North Africa and Europe. The merging together of all the continents into a single super-continent he called 'Pangaea'. Pangaea later 'drifted' or broke up into Gondwanaland and Laurasia, which finally split into the continents we find today.

Although Wegener and his supporters accumulated evidence, mainly from the fossil record on either side of the 'drifting' continents, to support continental drift, many scientists remained sceptical.

However, in the 1950s and early 1960s, through research in palaeomagnetism (ancient magnetism preserved in rocks) a revolution in thinking occurred. When molten material containing magnetic minerals cools and solidifies into rock, the direction of magnetisation in the rock is the same as that of the Earth's magnetic field at that particular time. The subject of the Earth's magnetism and palaeomagnetism will be discussed later in another context.

In the 1950s, geologists at several institutions worldwide,

including the Geological Survey of Canada, found evidence from different locations that indicated that the Earth's magnetic poles had moved all over the globe for hundreds of millions of years. By plotting the position of these poles, they came up with the phenomenon referred to as 'polar wandering'.

As it was considered unlikely that the magnetic poles had moved so dramatically in time, it was finally concluded that the continents had themselves moved, carrying the rocks with them.

The principles of moving plates or theory of 'plate tectonics' was established and Wegener was vindicated. Plate tectonics theorises that Earth's crust is composed of a series of large plates that 'float' on Earth's moving mantle. When these plates move, they can cause the creation of mountain ranges and other features such as earthquakes, faults and volcanoes as well as ocean basins and ridges and deep-sea trenches. The movement of whole continents is referred to as continental drift.

In plate tectonics, there are two types of plates: the thick continental plates and the relatively thin plates underlying the oceans. The movement of the plates is believed to be caused by convection currents in the mantle. When continental plates collide, they produce mountain ranges such as the Himalayas. When plates split apart, as on the ocean floors, magmatic material from the mantle moves up to fill the space. Other phenomena occur where continental plates meet oceanic plates and where plates slide past each other without a collision.

As a concrete example, there was an extraordinary discovery, reported in the journal *Nature* in April 2006. Researchers not only provided fossil evidence of an evolutionary 'missing link' between fish and land animals dating about 375 million years ago in what is now the Arctic wilderness about 600 miles from the North Pole, but established that the region had originally been an equatorial river delta before continental drift moved the land mass northward.

The Search for Alien Solar Systems

The tantalising search for solar systems similar to our own, possibly harbouring an Earth-like planet that is theoretically able

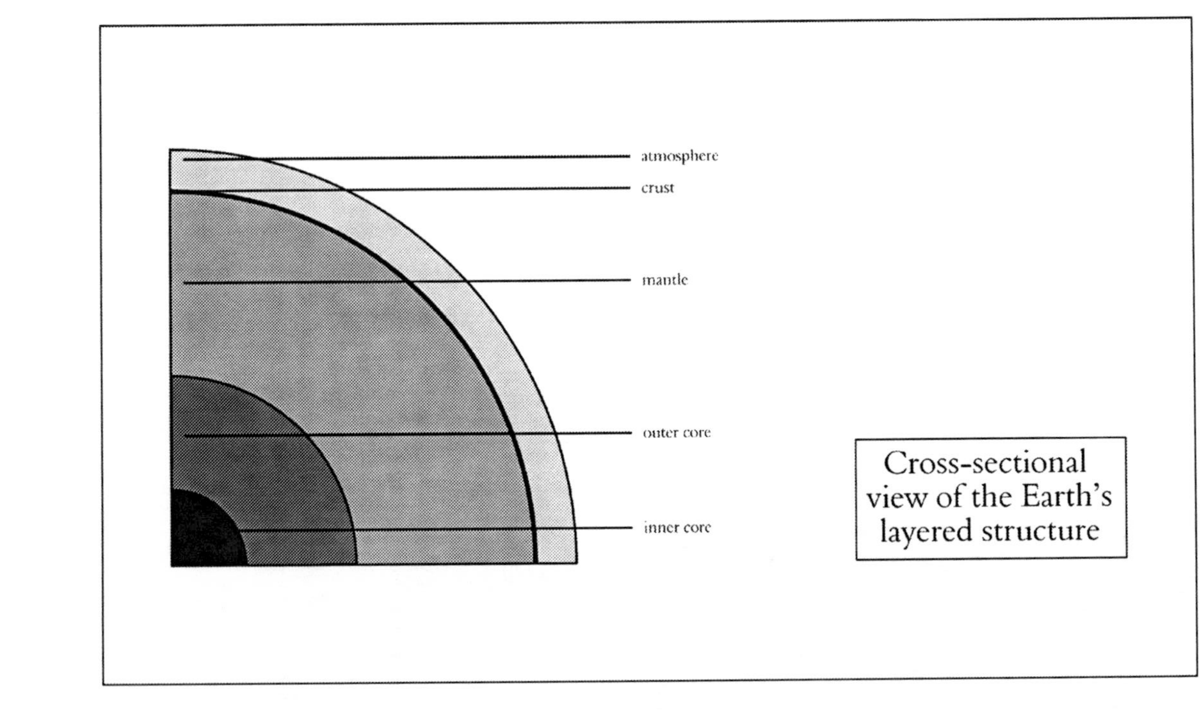

Cross-sectional view of the Earth's layered structure

to support living creatures, may some day bear fruit. Since 1996, numerous extra-solar planets have been identified.

However, in June 2002 a newly-reported gaseous planet four times larger than Jupiter was detected. It is reported to be about 512 million miles from a star known as 55 Cancri in the constellation Cancer. In comparison, in our own system, Jupiter is 484 million miles from our sun. This new planet takes thirteen years to circle its sun, only slightly longer than Jupiter's twelve-year orbit. Astronomers have indicated that the Jupiter-like object around 55 Cancri greatly improves the chances that an Earth-like planet will eventually be found.

Ongoing study of data from the Hubble Space Telescope has resulted in what may be the discovery of the oldest and most distant planet yet found in the universe. As reported by the Advertiser News Services (with contributions by the Associated Press and *Washington Post*) of 11 July 2003, astronomers have detected a planet, in the form of a gaseous sphere with more than twice the mass of Jupiter, believed to be 12.713 billion years old and 5,600 light years away. The possibility for the emergence and later disappearance of some form of life 13 billion years ago certainly has immense implications.

II
History of Life

Origin of Life

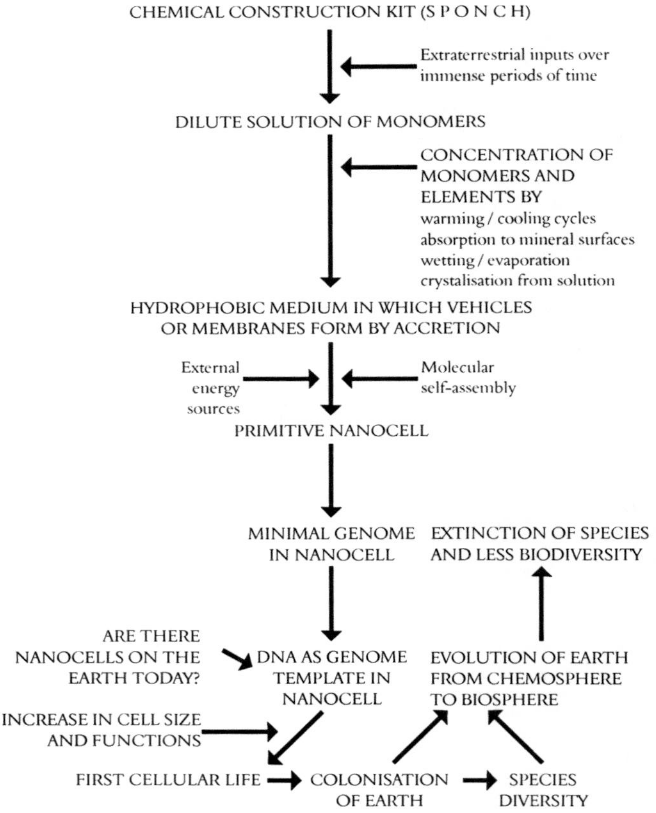

CHEMICAL CONSTRUCTION KIT (S P O N C H)

Extraterrestrial inputs over immense periods of time

DILUTE SOLUTION OF MONOMERS

CONCENTRATION OF MONOMERS AND ELEMENTS BY
warming / cooling cycles
absorption to mineral surfaces
wetting / evaporation
crystalisation from solution

HYDROPHOBIC MEDIUM IN WHICH VEHICLES OR MEMBRANES FORM BY ACCRETION

External energy sources

Molecular self-assembly

PRIMITIVE NANOCELL

MINIMAL GENOME IN NANOCELL

EXTINCTION OF SPECIES AND LESS BIODIVERSITY

ARE THERE NANOCELLS ON THE EARTH TODAY?

DNA AS GENOME TEMPLATE IN NANOCELL

EVOLUTION OF EARTH FROM CHEMOSPHERE TO BIOSPHERE

INCREASE IN CELL SIZE AND FUNCTIONS

FIRST CELLULAR LIFE → COLONISATION OF EARTH → SPECIES DIVERSITY

Proposed sequence of major events in the origin of a cell capable of growth, division and diversification.

Source: Jack Trevors and Roland Psenner, FEMS Microbiology Reviews vol. 25 (2001) pp.573 – 582

Primitive Life

From the universe's awesome beginning, which may very well be a repeat of a prior or even a parallel big bang in a multiverse, we have all the components for the development of biological processes on Earth. Microbiologists have made great strides in researching early bacterial molecular evolution. The proposed sequence of major events for the self-assembly of life on Earth has been detailed in a compelling paper by Trevors and Psenner.*

From simple elements and components in a lifeless chemosphere the first cell capable of growth and division is considered to have assembled in a reducing, anoxic environment of hydrogen, ammonia and methane. Energy sources, early proteins or enzymes and subsequent DNA and RNA were part of the assembly of life process. Gases such as hydrogen, nitrogen, methane and carbon dioxide were present on the early Earth and are considered to have important roles in bacterial metabolism. Hydrogen is also believed to be a universal energy source that the cells were capable of using during the Earth's early history. The proposed sequence of major events in the origin of a cell capable of growth, division and diversification is summarised in the figure shown opposite. The absolute link between primitive nanobacteria and present-day bacteria is currently under investigation. Nevertheless, convincing support for the theoretical models of the early chemistry of solar systems where large amounts of complex molecules were predicted to be found in regions surrounding an infant star was confirmed jointly by astronomers at the Leiden Observatory in the Netherlands as well as the W M Keck Observatory on Mauna Kea, Hawaii, and the Jet Propulsion Laboratory of the California Institute of Technology on 21 January 2005.

Support for input of organic matter from outer space to our Earth was reported in the journal *Science* of 30 November 2006.

---

* Trevors and Psenner, *Federation of European Microbiological Societies Microbiology Reviews*, 2001, vol. 25, pp.573–582.

Fragments of a meteorite that crashed onto frozen Tagish Lake on the British Columbia–Yukon border in 2000 have been analysed and found to contain tiny, hollow carbon spheres mixed with hydrogen and nitrogen. Also, according to the study, the meteorite is made up of materials that may be more than 4.5 billion years old, which is older than the Earth, other planets, and even the sun. The meteorite is believed to have travelled approximately 22 million km before landing.

The disc of hot gases surrounding an infant star (referred to as IRS 46 and located about 375 light years from Earth) was found to be rich in molecules of acetylene and hydrogen cyanide. Combined with water and under the right conditions, these molecules can interact to produce a variety of organic compounds such as amino acids and adenine. Adenine is a component of both DNA and RNA, two nucleic acids described as the 'building blocks' of living organisms.

Cells are the structural and functional units of life, and the manner in which they self-replicate and differentiate in the development of individuals is becoming generally understood. Events within and between cells can be explained in terms of various chemical reactions. In molecular biology, knowledge of the genetic make-up of individuals within the same or different species is made available for a better understanding of their relationships.

In embryology, or the study of the embryo, we find that all living things pass though a continuum from the least complex to their final complexity. They go through what is called *recapitulation*, where even the embryos of mammals, including humans, at early states of development, go through very similar morphological phases found in earlier forms of life in the evolutionary patterns of development. In a similar manner, a common plan for the development of the nervous system, the kidneys, liver, lungs, heart and brain development can be observed.

The chromosomes form the cell nucleus of multicellular organisms and contain the DNA or RNA that comprise the genes of the individual. In sexual reproduction, each of the sexes contributes half of the chromosomes. For instance, in humans there are forty-six chromosomes, twenty-three of which are contributed by the male and twenty-three contributed by the female. More specifically, the twenty-three pairs of chromosomes comprise twenty-two autosomes, plus the X/Y chromosome. Our close relatives, the chimpanzees, gorillas and orang-utans each have forty-eight chromosomes, which suggests that at some time in our ancestry two pairs of chromosomes somehow fused together.

The number of chromosomes varies widely in different organisms. It is reported that the record for the minimum number of chromosomes belongs to a subspecies of the ant, *myrmecia pilosula*, in which females have a single pair of chromosomes. This species reproduces by a process called haplodiploidy, in which fertilised eggs (diploid) become females, while unfertilised eggs (haploid) develop into males. The males in this group of ants have in each of their cells a single chromosome. The record maximum number of chromosomes is found in a species of fern. There are approximately 630 pairs of chromosomes or a total of 1,260 chromosomes per cell. Other examples include the mosquito which has six chromosomes, a bean with twenty-two, a goat with sixty, a dog with seventy-eight and a king crab with two hundred and eight. It would be convenient if there were some type of 'evolutionary ladder', where more complex organisms have more chromosomes, but this is not the case.

The lifespan of a chromosome is one generation. However, genes travel intact from grandparent to grandchild and are considered to be close to immortal.

When death occurs in an individual, the brain and organs stop functioning, but cells continue to replicate for a period of time as evidenced by hair and nail growth. However, at later stages where decomposition is initiated, cells desynthesise eventually into the basic organic or inorganic compounds they were synthesised from.

Recent discoveries of spherical and ovoid structures representing primitive micro-organisms in rocks and minerals of the Australian outback are now believed to date 'life' on Earth as far back as 3.47 billion years ago.

In the past few years, researchers have found just how diverse and widespread life can be on Earth. As a recent example, researchers have found bacteria in a hole more than 4,000 feet in volcanic rock on the Big Island of Hawaii. The study was funded by NASA, the Jet Propulsion Laboratory, California Institute of Technology and OSU. The results were published in the December issue of the journal *Geochemistry, Geophysics and Geosystems*. Other examples of microbes found thriving in what used to be considered inhospitable environments include boiling hot springs and even nuclear reactors.

From these oldest branches on the tree of life, evolutionary processes were able to develop all the varied kingdoms of life. For the animal kingdom in particular, recent genetic investigations substantiate earlier studies that the vast majority of animals belong to one of three primary evolutionary lines.

As the animal fossil record stretches back only 540 million years, genetic analysis could reveal something about the genes of the common ancestor of all animals that lived somewhat more than 600 million years ago. It is noteworthy that at about the fifth century BCE, the Ajivikas, who were for some centuries the primary rivals to both Buddhism and Jainism, taught that every soul must traverse 8,400,000 lives. Considering a lifespan of sixty-five years, this would – amazingly – coincide with the oldest animal fossil record.

Palaeoanthropology

Coming now to the relationship of humans to other animals, palaeoanthropologists have acquired most of their knowledge through the study of fossilised remains and by placing each fossil correctly on a timeline.

Humans have been found to belong to the primate class of mammals, which includes apes and which appeared more than 60 million years ago. From subgroups to subclasses we come to the genus of hominids that evolved about 5 million years ago. In 1971, a skeleton called Lucy was found in Ethiopia that was dated to have lived between 3.5 and 2.8 million years ago and had anatomical characteristics about halfway between those of apes and humans. Lucy was believed to be modern humans' direct antecedent. However, the discovery of an unusual 3.5 million-year-old skull in northern Kenya in 1998–1999, known as *Kenyanthropus platyops*, may be another candidate. More recently discovered in Ethiopia, and reported in the journal *Nature* and prominently featured in *Time**, were the remains of what appears to be the most ancient ancestor ever discovered. The chimp-size creature, named *Ardipithecus ramidus kadabba*, is believed to have lived between 5.8 million and 5.2 million years ago. It is thought that from the shape of the mid-foot bones that *kadabba* almost certainly walked upright most of the time.

A press release of 10 July 2002, noted a discovery in the Djurab Desert in northern Chad by a French-Chadian team: fossils of a skull, two fragments of a lower jaw and three teeth – from at least five individuals the size of chimpanzees – were exposed by a sandstorm. The fossils are considered to be from a previously unknown genus and species that may very well stand at the very base of the human family tree. This ancestor lived 6–7 million years ago and is called *Sahelanthropus Tchadensis*, nicknamed 'Toumai' or 'hope of life'.

A newly-discovered ape species, *Pierolapithecus catalaunicus*, discovered near Barcelona, Spain, has recently electrified scientists. In the 18 November 2004 issue of *Science*, the species is dated at almost 13 million years old and is speculated to be the latest probable common ancestor to all living humans and great apes. However, as gaps in the fossil record are still present, even more candidates may be found. A change to a more modern form of *Homo sapiens* or 'the knowledgeable one' is believed to have originated in Africa within the past 200,000 years.

The early hominids exited Africa to Asia and other parts of the Old World in successive waves starting at about 1.5–2 million years ago. The question of whether humans evolved simultaneously on a multi-regional basis, or whether all modern humans – *Homo sapiens* – are descendents of the last wave of migration estimated at 100,000–120,000 years ago, is still being debated. The multi-regionalists base their evidence mainly on anatomical studies on fossil remains and associated artefacts. The common ancestor proponents base their results mainly on genetic studies.

---

* Lemonick and Dorfman, 'The Dawn of Man', *Time* magazine, 23 July 2001, vol. 158, no. 3.

In this regard, molecular biologists have found that mitochondrial DNA determinations provide the basis for a reliable molecular clock. Mitochondrial DNA encodes only thirty-seven genes and is inherited from the mother alone. Using this technique, the maternal lineages of people around the world may be traced to a common ancestor.

Early humans are considered to be just like present-day humans in basic character and intelligence. The difference lies only in cultural style and in the greater organisational complexity of the world today. From magnificent sketches, engravings and paintings, discovered in 1868 in caves in France, we have learned that these hunting and gathering societies made tools, including needles and fish hooks and other implements. Radiocarbon dating indicates that this artwork was done between 17,000 and 20,000 years ago. Other carved figures found in a cave in southwest Germany date from about 22,000 years ago. More recent archaeological studies in Africa, Australia, Europe and the Middle East have uncovered carvings, weapons and paintings that suggest modern humans evolved earlier than previously thought. Ivory beads and other artefacts dating to approximately 100,000 years ago have been discovered in Africa. Rock paintings dating back about 60,000 years have been found in Australia, and a burial cave in Israel has been dated to about 100,000 years ago.

Genetic studies by biologist Alan Templeton of Washington University in St Louis seem to indicate that Neanderthals and other early branches of our family tree were not evolutionary dead ends. On the contrary, anthropologist John Relethford at the State University of New York, Oneonta, interpreted this to mean that '...instead of replacing existing people, they mixed with other groups'.

These ancestors of ours had the same instincts, feelings and imagination. They realised their great dependence on nature, especially on the various animals and plants in their local environment. Other ingredients included the need for group and cultural activities. Cooperative effort was vital for each group's survival.

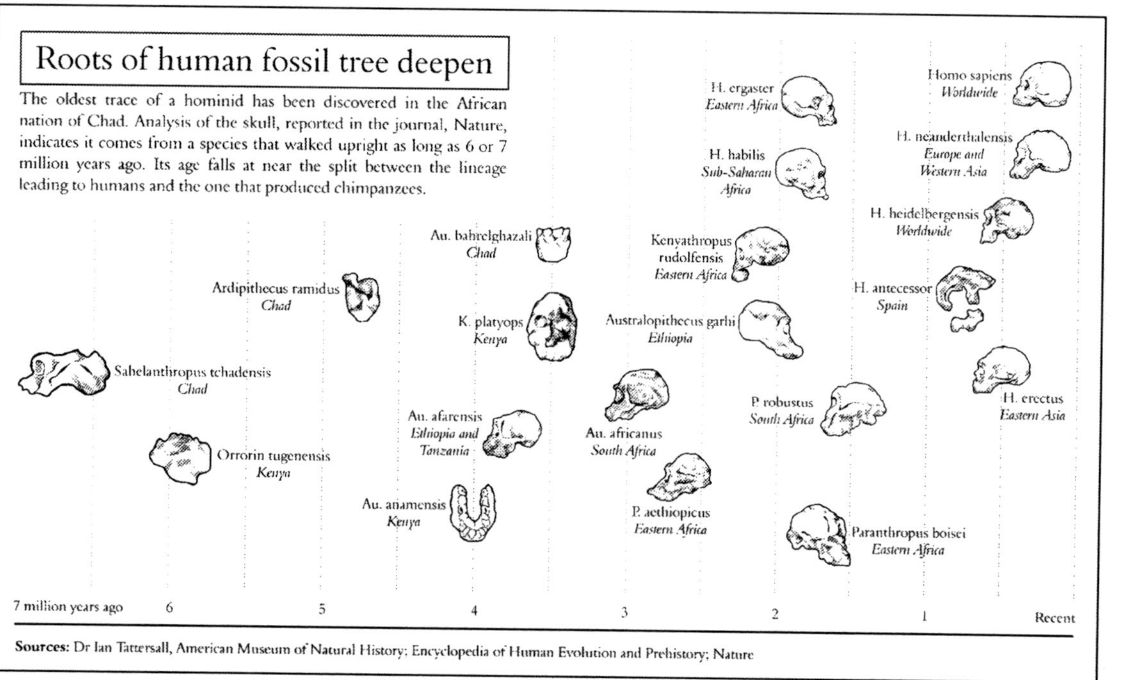
Roots of human fossil tree deepen
The oldest trace of a hominid has been discovered in the African nation of Chad. Analysis of the skull, reported in the journal, Nature, indicates it comes from a species that walked upright as long as 6 or 7 million years ago. Its age falls at near the split between the lineage leading to humans and the one that produced chimpanzees.
Sahelanthropus tchadensis
Chad
Orrorin tugenensis
Kenya
Ardipithecus ramidus
Chad
Au. bahrelghazali
Chad
K. platyops
Kenya
Au. afarensis
Ethiopia and Tanzania
Au. anamensis
Kenya
Au. africanus
South Africa
P. aethiopicus
Eastern Africa
Kenyathropus rudolfensis
Eastern Africa
Australopithecus garhi
Ethiopia
H. ergaster
Eastern Africa
H. habilis
Sub-Saharan Africa
P. robustus
South Africa
Paranthropus boisei
Eastern Africa
Homo sapiens
Worldwide
H. neanderthalensis
Europe and Western Asia
H. heidelbergensis
Worldwide
H. antecessor
Spain
H. erectus
Eastern Asia
7 million years ago
6
5
4
3
2
1
Recent
Sources: Dr Ian Tattersall, American Museum of Natural History; Encyclopedia of Human Evolution and Prehistory; Nature

Communication and Linguistics

Man's early communication was by bodily gestures and primitive utterances. The earliest sounds were probably merely reflex, involuntary and independent of the will, produced under the impulse of the emotions. Gestures, such as blinking the eyes or shrugging the shoulders, and primitive sounds, were originally involuntary expressions of emotion, confined at first to the individual, but as the need for intercommunication was more strongly felt, they gradually reached a high state of development.

More developed means of communication and expression of ideas can be found in traditional Egyptian picture writing. In Egyptian texts they are called 'the words of the god', and are referred to by classical writers as 'hieroglyphic' i.e. 'sacred carving', and date back to 4,000 BCE.

Deciphering of the Egyptian hieroglyphs was accomplished by studying the Rosetta Stone (now in the British Museum) and a stone obelisk from Philae (now stands in the park at Kingston Lacy, Dorset). Each of these monuments contains a Greek as well as an Egyptian version of the inscription. It was found that every hieroglyph can be used to express an idea, in which case it is called an 'ideograph', or a sound, in which case it is called a 'phonetic'. Phonetics were found to be either alphabetic or syllabic. J F Champollion (1790–1832) has been given major credit for the decipherment and who is noted for publishing a *Hieroglyphic Grammar and Dictionary*.

The oldest living language is undoubtedly Chinese, which was spoken at least 5,000 years ago on the banks of the Yellow River, and has survived with comparable little change to the present day. Chinese is monosyllabic and uninflected, that is to say, every word is expressed by a single syllable which is unaffected by number, person, mood, etc. In the official Mandarin dialect there are 400 syllables, so that the same sound has to do duty for many different words. This would of course lead to great ambiguity, were it not that each word has its own 'tone' or modulation of the voice. There are four tones. The written language is composed of signs or characters and makes its appeal to the eye rather than to the ear. The origin of the Chinese script was pictorial, and a few of the existing characters are still recognisable as slightly conventionalised pictures. These are called pictograms. A further step was taken by the formation of ideograms – a compound of two or more pictures, suggesting an idea. And lastly, there are phonograms – compound characters in which one element (the phonetic) represents only the sound, or an approximation to the sound, of the whole, while the other (the radical) gives a general clue to the sense by indicating the category of things to which the word belongs.

The Rosetta Stone

Other languages, or rather groups of languages, including Indo-European, Ural-Altaic, and Semitic, as well as the indigenous languages of North and South America, Africa, Asia and Australia have had long histories of development. With regard to the Indo-European family of languages, Russell Gray,* an evolutionary biologist at the University of Auckland, New Zealand, co-authored a paper published in *Nature* in November 2003, which showed that the ancestral tongue known as proto-Indo-European existed about 8,700 years ago, making it considerably older than what linguists have previously assumed. Many researchers believed that the Indo-

* Gray and Atkinson, 'Language-tree divergence times support the Anatolian theory of Indo-European origin', *Nature*, 27 November 2003, vol. 426, no. 6965, pp.435–438

European languages were spread around 6,000 years ago by the Kurgan warriors who initially resided in the Russian steppes. The rival theory holds that the Indo-Europeans were the first farmers who lived in ancient Turkey and spread their language some 8,000 years ago through new agricultural techniques and not through conquest. Both of these theories are depicted in the diagram below.

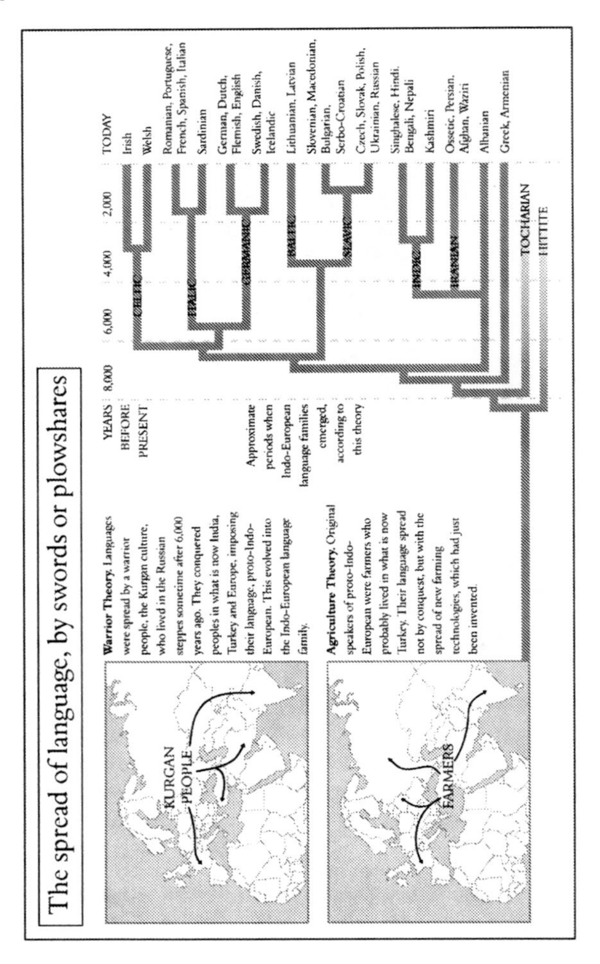

These language groups evolved from relatively simple forms of communication through the interaction of different dialects and languages. Alphabets, grammar, level of inflection, consonantal combinations and so on were incorporated and adapted in one form or another. It was not until about 1,500 BCE that the first true alphabet was invented in Canaan, in the coastal region of the Mediterranean occupied by modern-day Syria, Lebanon, Palestine and Israel. As in other branches of evolution, some languages thrived while others became extinct. Currently, there are about 6,000 languages worldwide, with the highest concentration in Papua New Guinea, home to at least 820 different tongues. According to Michael Krauss of the University of Alaska at Fairbanks, the number of languages and/or dialects spoken 10,000 years ago, when the total human population was approximately 10 million, may have been as many as 20,000.

Ancient Greek and Latin will be used as models to exemplify the unique qualities and adaptivity of languages. In the example of Ancient Greek or Attic, the language was perfected as a reasoning instrument. It was highly suitable for rendering abstruse and scientific terms. The Greek philosopher-mathematician, Thales, realised the importance of organising mathematics upon a logical basis. The tradition was further developed by Pythagoras, Plato and Aristotle. In a later period, we have the distinguished mathematicians such as Euclid, Archimedes, Apollonius, Ptolemy and others.

The Latin language, on the other hand, was considered less poetical and less imaginative than Greek, but extremely practical. Latin was considered eminently adapted for legal, historical, religious and political compositions which demanded dignity and sonority for their successful execution.

Languages have had and still have a profound effect on culture, which in turn has a reciprocal influence in the refinement of language. The extension of language and logic has flowed into mathematics. Other expressions of the mind are found in art, sculpture and music.

The question of the origin of language has been the subject of discussion from the earliest times. After the introduction of Christianity, language was generally regarded as a divine revelation, a theory revived as late as the nineteenth century by L Bonald (1754–1840), who held that speech preceded thought and that therefore it must have been directly communicated by God.

Philosophy, Mathematics and Logic

From the time of Aristotle until the nineteenth century, logic had come to be thought of as consisting of laws that govern thought. It was not until the nineteenth and twentieth centuries that revolutionary thinking by mathematician-philosophers overthrew this conception.

It was the German philosopher and mathematician Gottlob Frege (1848–1925)* who stated that logic is objective and that logical propositions are objective truths. He made it clear that philosophy ought to be logic-based and not founded on the psychology of human beings. He was in the forefront of developing symbolic logic.

Frege also used logic in the understanding of mathematics. Until his time, German philosophers believed that mathematics was a product of the human mind. By proposing that mathematics was a type of language, he was able to explain mathematics by way of logical operations. In 1879 he published a pamphlet describing a new calculus, which has since been at the centre of modern logic.

The eminent British mathematician-philosopher Bertrand Russell (1872–1970) also arrived at the conclusion, independently of Frege, that mathematics could be derived from the fundamental principles of logic. In collaboration with Alfred North Whitehead, and using Frege's groundwork as well as his own, they produced *Principia Mathematica*† regarded by many as the greatest contribution to logic since Aristotle.

The Austrian philosopher-mathematician, Ludwig Wittgenstein

---

* Frege, *Begriffschrift* (Concept Script), second edition, Hildescheim, G Olm, 1964 (1879).

† Whitehead and Russell, *Principia Mathematica*, second edition, London, Cambridge University Press, 1935.

(1889–1951)* studied philosophy under the guidance of Russell. In his earlier work, he was convinced, like Frege and Russell, that language was based on logic. He also believed that philosophical language should be restricted to observable objects and not be used to make assertions about topics such as art, ethics, metaphysics and religion.

In later life, he dramatically modified his views and came to believe that what was most unsatisfactory in his earlier philosophy was its so-called 'picture' theory of meaning. In other words, a painter can represent on a small piece of canvas the whole countryside, just as words in sentences can describe reality accurately or, for that matter, inaccurately in language. For these relationships, he used the term 'logical form'. However, language can carry out many other tasks besides the picturing of reality. Wittgenstein dropped the picture metaphor and adopted instead the metaphor of a tool. As a tool, language can be used to carry out an indefinite amount of different undertakings. He believed that languages are fluid, and that specific words may not always have specific meanings. He claimed that language is public, and is learned and used in a social context. He believed that words acquired their meanings from the activity or endeavour involved. Scientific activity, religious activity, musical activity, etc. have their own language and derive their meanings from the ways they are used in their separate worlds. Linguistic philosophy reached its peak in the twentieth century and has a considerable influence in current thinking.

* Wittgenstein, *Philosophical Investigations*, third edition, translated by G E M Anscombe, edited by Anscombe and Rhees, New York, Macmillan, 1958.

III
Religion

Natural Phenomena – Awareness and Reverence

Our ancestors were exposed to and were probably more influenced by natural phenomena than we are today. Lightning, thunder, hailstorms, volcanic eruptions and other phenomena were feared and respected. The sun, moon and planets were revered and their rhythmic relations to earthly cycles were compared.

Our ancestors were fascinated by the mysteries of birth, life and death of not only humans but of the whole animal kingdom. They became aware of 'the wonder and bloom of the world'.

To express their reverence to the haunting mystery and beauty of the universe, they directed their piety to numerous gods – the sun god, the moon god, the star god and unaccountable others. In fact, in Indian classical culture there are traditionally said to be 330 million personal and mythic gods.

In ancient Egypt, dating back to beyond 3,000 BCE, there was much concern with the nature of death and the quest for survival after death. The building of the Sphinx and pyramids and the technology of mummification haunt our imagination.

With the ongoing search for answers to the eternal mystery, certain ideas and rituals were consecrated and others modified. The patterns of public and private behaviour became codified. The theme of dying and rising gods played a constant role in daily life, and may have influenced later thinking.

Cultural Evolution

One of the unique attributes of man is his culture, which has evolved over the last few thousand years. Cultural evolution includes examples such as language, art, customs, architecture, and religious beliefs. Although this cultural evolution has played and continues to play a significant role in human evolution, culture is not transmitted from generation to generation from our gene pool. The catch-term used for cultural transmission is referred to as the 'meme', coined by Richard Dawkins in his book entitled *The Selfish Gene*. The survival value of a meme is dependant on its long-term acceptability among the population at large. Popular songs and melodies may be all the rage for the moment but they only have a short shelf life. Scientific ideas are passed on from generation to generation in scientific journals, and sometimes with a few changes become brilliant advances in comprehending the many forces of nature and of the universe as a whole. In the case of religious beliefs such as the Judaeo-Christian and other beliefs, they have survived in written and oral form for several millennia.

Religions of Semitic Origin

In the Middle East, the Israelites, a nomadic desert tribe, made a significant contribution to religion and ethics. It was in their affirmation of monotheism that a galaxy of geniuses – Abraham, Joseph, Moses, Samuel, Isaiah, the Unknown Isaiah, Jeremiah and Ezekiel – 'arose before us and as prophets burned'. The fervour of this new religious ideology with strict commitment and loyalty to one God, and no other God, provided sustenance through periods of oppression and persecution.

The spiritual momentum never being quiescent, there soon burst forth another glorious three: Yehoshua (Jesus), Yohanan his cousin (John the Baptist) and Shaoul (Paul) – all devout Jews. These stellar intellects planted the standards of religion and advanced ethics to an even higher plane. It was Jesus whose sense of the mystery, the other world and love of the Creator allowed him to teach a higher set of moral values through parables and stories. It was Paul, teaching in Greek, who transmitted this inspiring modified religion, first known as New Judaism, to cities in Greece, Asia Minor and to Rome itself. Easier access to Gentile converts was made by relaxing or eliminating ritual restrictions such as the solemn rites of circumcision and strict observance of dietary laws. The spread of Christianity in all its diverse forms is a testament to its universal appeal.

Some 600 years later, another stellar intellect was born in Mecca. At an early age, he had revelations and striking visionary experiences. At the age of forty, the Prophet Muhammad decided to devote himself to being a messenger of Allah, the one and only God, and creator of the universe. The Qur'an, Islam's holy scripture, is acknowledged as God's eternal speech. Muslims, like Christians and Jews, consider themselves to be direct descendants of the Prophet Abraham, and believe that Muhammad was the last prophet in the great line of prophets.

Buddhism

In the sixth century BCE, a surge of new ideas and practices was occurring in a region centred on the Ganges. Siddhartha Gautama (the Buddha or the Enlightened One) worked tirelessly for the sake of other human beings. The concepts of rebirth or reincarnation and liberation from an unsatisfactory world were addressed. A system of meditation with understanding was developed. Ethical conduct, trust and compassion were important. The impermanence and interdependence of everything was one of the central insights. Love and compassion for animals and other forms of life was stressed. The Buddha took the moderate middle road regarding whether the soul was eternal or cut off at death.

Hinduism

Hinduism is considered to be one of the world's most ancient religions and is believed to have developed over a period of about 4,000 years, originating in the Indus Valley region of India.

Contrary to other religions, Hinduism has no founder but is regarded by its adherents as having existed for ever. It is both a religious as well as a cultural system, embracing and assimilating a broad range of philosophies from monotheism to polytheism. The concepts of karma (action, deed), dharma (that which contains or upholds the cosmos), reincarnation, temple worship, manifold deities and an all-pervasive divinity are part of the vast heritage of culture and belief. Spirituality, enlightenment, revelation and transcendental meditation play important roles. In addition, there are Puranas, or folk narratives, containing ethical and cosmological teachings relative to God, man and the world. They revolve around five subjects: primary creation, secondary creation, genealogy, cycles of time and history.

In an unusual book by Satguru Sivaya Subramuniyaswami entitled *Lemurian Scrolls, Angelic Prophecies Revealing Human Origins*, the Hindu concept of cosmic cycles within the infinitely recurring periods of the universe is given credibility. Drawing on the lengths of the major time periods of the Hindu system and reflecting calculations given in the Scrolls, a timeline comparison with scientific data is presented. Examples include the following:

TIMELINE

435 million years

Scrolls Souls come to Earth and incarnate as animals.

Science First evidence of plant life.

410 million years

Science First land animals appear.

4,318,121 years

Scrolls Celestial beings manifest Earth bodies around their subtle form.

Science 'Human' life appears.

It is to be noted that recent palaeontological evidence indicates that plant life may have originated on land almost 475 million years ago or some 40–50 million years before the appearance of land animals.

Religious Development

In the formation and development of religious tradition, the key is *religare*, the binding together of people to a certain belief. They include, to a greater or lesser degree, the following: sensations of hope; love and gratitude; ritual; meditation; mysticism; visions; sacrifices; heroes, martyrs and saints; souls; divine cloning; documents and true scriptures. These sensations are fortified by the awe-inspiring structures of temples and cathedrals.

The dominant factor in their survival includes a strong moral code and some force or deity that can, at will, nullify the laws of nature. The dogma of a faith is based on pronouncements and revelations of prophets and priests. Examples of miracles in the Old and New Testaments are attributed to the direct intervention of a supernatural force or person referred to as God. This dogma was interpreted as divinely inspired and preached by the learned class of the day – the priests – and received and accepted by the population as a whole. This common acceptance by society in general becomes a very formidable unifying force.

For a religious system to be successful in 'universalising'* it must:

- be considered by its adherents to be proper to all mankind;
- have mechanisms to facilitate its transmittal;
- have at some time successfully broken through the restrictions of a special relationship to place or particular social group; and
- have been established as a dominant religion at least on a regional scale.

Buddhism, Christianity and Islam are the three major universalising systems. They are 'religions of revelations', and began as revelations in a closed ethnic field. The message of Jesus Christ was communicated specifically to the Jews; that of Muhammad initially circulated chiefly among the Arabs of Central Arabia.

---

* Sopher, D, *Geography of Religions*, New Jersey, Prentice Hall Inc., 1967.

Using techniques such as simple conversion and missionary zeal has permitted the rapid expansion of these systems. Admission is permitted readily in contrast to the long process of acculturation and absorption into the ethnic systems. In the case of Islam, adherence is demonstrated by the recital of a formula. In Buddhism, adherence is a simple voluntary act. In Christian communities, the rites of baptism and communion constitute admission.

However, as comforting and convincing as these religious cultures may be, a noted religionist has cautioned:

> The stars remain but astronomies change.

This is interpreted to mean that the universe is eternal and that our religious conceptions are continually being modified to suit the changes in our knowledge. In the same manner astronomies must be adapted to fit new facts discovered or revealed by increasingly refined instruments at our disposal.

As we have seen, religious beliefs are not determined by heredity but are learned or indoctrinated within specific social and/or environmental situations.

In all these venues, religion can be visualised as the 'missing chip' in the human supercomputer. The ultimate goal would be to provide an 'absolute unifying chip' comparable in essence to the proposal in the 'absolute unifying theory' defined in part V.

Catalogue of Religions

A comprehensive list of adherents of all religions by six continental areas is catalogued in the following table:

The 2004 Annual Megacensus of Religions

Worldwide Adherents of All Religions, mid-2004

Table Notes

Continents. These follow current UN demographic terminology, which now divides the world into the six major areas shown in the table. *See* United Nations, *World Population Prospects: The 2002 Revision* (New York: UN, 2003), with populations of all continents, regions, and countries covering the period 1950–2050, with 100 variables for every country each year. Note that 'Asia' includes the former Soviet Central Asian states, and 'Europe' includes all of Russia eastward of the Pacific.

Countries. The last column enumerates sovereign and non-sovereign countries in which each religion or religious grouping has a numerically significant and organised following.

Adherents. As defined in the 1948 Universal Declaration of Human Rights, a person's religion is what he or she professes, confesses, or states that it is. Totals are enumerated for each of the world's 238 countries following the methodology of the *World Christian Encyclopaedia*, second edition, 2001, and *World Christian Trends*, 2001, using recent censuses, polls, surveys, yearbooks, reports, websites, literature, and other data. See the World Christian Database *www.worldchristiandatabase.org* for more detail. Religions are ranked in order of worldwide size in mid-2004.

Christians. Followers of Jesus Christ, enumerated here under **Affiliated Christians**, those affiliated with churches (church members, with names written on church rolls, usually total number of baptized persons including children baptized, dedicated, or undedicated): total in 2004 is 1,998,631,000, shown above divided among the six standardized ecclesiastical blocs and with (negative and italicised) figures for those persons with **Multiple Affiliation** (all who are baptized members of more than one denomination); and **Unaffiliated Christians**, who are persons professing or confessing in censuses or polls to be Christians though not affiliated.

Independents. This term here denotes members of Christian churches and networks who regard themselves as postdenomenationalist and neo-apostolic and thus independent of historic, mainstream, organised, institutionalised, confessional, denominationalist Christianity.

Marginal Christians. Members of denominations who define themselves as Christians but who are on the margins of organised mainstream Christianity (e.g. Unitarians, Mormons, Jehovah's Witnesses, Christian Scientists and Religious Scientists).

Muslims. 83% Sunnites, 16% Shi'ites, 1% other schools.

Hindus. 70% Vaishnavites, 25% Shaivites, 2% neo-Hindus and reform Hindus.

Nonreligious. Persons professing no religion, non-believers, agnostics, freethinkers, uninterested, or dereligionised secularists indifferent to all religion but not militantly so.

Chinese universists. Followers of a unique complex of beliefs and practices that may include: universism (yin/yang cosmology, with dualities Earth/Heaven, evil/good, darkness/light), ancestor cult, Confucian ethics, divination, festivals, folk religion, goddess worship, household gods, local deities, mediums, metaphysics, monasteries, neo-Confuscianism, popular religion, sacrifices, shamans, spirit writing, and Taoist and Buddhist elements.

Buddhists. 56% Mahayana, 38% Theravada (Hinayana), 6% Tantrayana (Lamaism).

Ethnoreligionists. Followers of local, tribal, animistic, or shamanistic religions, with members restricted to one ethnic group.

Atheists. Persons professing atheism, scepticism, disbelief, or irreligion, including the militantly anti-religious (opposed to all religion).

Nonreligionists. Followers of Asian twentieth-century neo-religions, neo-religious movements, radical new crisis religions, and non-Christian syncretistic mass religions.

Jews. Adherents of Judaism. For detailed date of 'core' Jewish population, see the annual 'World Jewish Populations' article in the American Jewish Committee's *American Jewish Year Book*.

Confucianists. Non-Chinese followers of Confucius and Confucianism, mostly Koreans in Korea.

Other religionists. Including a handful of religions, quasi-religions, pseudo-religions, parareligions, religious or mystic systems, and religious and semi-religious brotherhoods of numerous varieties.

Total population. UN medium variant figures for mid-2004, as given in *World Population Prospects: The 2002 Revision.*

	AFRICA	ASIA	EUROPE	LATIN AMERICA	NORTHERN AMERICA	OCEANIA	WORLD	%	NO. OF COUNTRIES
Christians	401,717,000	341,337,000	553,689,000	510,131,000	273,941,000	26,962,000	2,106,962,000	33.0	238
Affiliated Christians	380,265,000	335,602,000	531,267,000	504,747,000	223,994,000	21,994,000	1,997,869,000	31.3	238
Roman Catholics	143,065,000	121,618,000	276,739,000	476,699,000	79,217,000	8,470,000	1,105,808,000	17.3	235
Independents	87,913,000	176,516,000	24,445,000	44,810,000	81,138,000	1,719,000	416,541,000	6.5	221
Protestants	115,276,000	56,512,000	70,908,000	53,572,000	65,881,000	7,699,000	369,848,000	5.8	232
Orthodox	37,989,000	13,240,000	158,974,000	848,000	6,620,000	756,000	218,427,000	3.4	134
Anglicans	43,404,000	733,000	25,727,000	909,000	2,986,000	4,986,000	78,745,000	1.2	163
Marginal Christians	3,269,000	3,083,000	4,425,000	10,352,000	11,384,000	630,000	33,143,000	0.5	215
Multiple Affiliation	*-50,562,000*	*-34,528,000*	*-10,021,000*	*-80,962,000*	*-23,217,000*	*-2,252,000*	*-201,542,000*	*-3.2*	*163*
Unaffiliated Christians	21,437,000	5,734,000	22,395,000	5,384,000	49,947,000	4,153,000	109,050,000	1.7	232
Muslims	350,453,000	892,440,000	33,290,000	1,724,000	5,109,000	408,000	1,283,424,000	20.1	206
Hindus	2,604,000	844,593,000	1,467,000	766,000	1,444,000	417,000	851,291,000	13.3	116
Chinese Universists	35,400	400,718,000	266,000	200,000	713,000	133,000	402,065,000	6.3	94
Buddhists	148,000	369,394,000	1,643,000	699,000	3,063,000	493,000	375,440,000	5.9	130

Ethno religionists	105,251,000	141,589,000	1,238,000	3,109,000	1,263,000	319,000	252,769,000	4.0	144
Neo religionists	112,000	104,352,000	381,000	764,000	1,561,000	84,800	107,255,000	1.7	107
Sikhs	58,400	24,085,000	238,000	0	583,000	24,800	24,989,000	0.4	34
Jews	224,000	5,317,000	1,985,000	1,206,000	6,154,000	104,000	14,990,000	0.2	134
Spiritists	3,100	2,000	135,000	12,575,000	160,000	7,300	12,882,000	0.2	56
Baha'is	1,929,000	3,639,000	146,000	813,000	847,000	122,000	7,496,000	0.1	218
Confucianists	300	6,379,000	16,600	800	0	50,600	6,447,000	0.1	16
Jains	74,900	4,436,000	0	0	7,500	700	4,519,000	0.1	11
Shintoists	0	2,717,000	0	7,200	60,000	0	2,784,000	0.0	8
Taoists	0	2,702,000	0	0	11,900	0	2,714,000	0.0	5
Zoroastrians	900	2,429,000	89,900	0	81,600	3,200	2,605,000	0.0	23
Other religionists	75,000	68,000	257,500	105,000	650,000	10,000	1,166,000	0.0	78
Nonreligious	5,912,000	601,478,000	108,674,000	15,939,000	31,286,000	3,894,600	767,184,000	12.0	237
Atheists	585,000	112,870,000	22,048,000	2,756,000	1,997,000	400,000	150,656,000	2.4	219
Total population	869,183,000	3,870,545,000	725,564,000	550,795,000	328,932,000	32,619,000	6,377,643,000	100.00	238

IV
Research

Scientific Investigation – Catastrophism vs Uniformitarianism

Research in the fascinating fields of geology, chemistry, physics, mathematics and biology have contributed significantly to our understanding of the Earth's physical features and life processes.

Before the eighteenth century, most Western scientists were of the view that the age of the Earth was just a few thousand years, as envisioned in the Bible. As this constraint did not provide sufficient time for the gradual occurrence of dramatic geological features such as mountain ranges and canyons, the theory known as Catastrophism was invoked.

Assuming a much older Earth, the theory of Uniformitarianism was introduced by the Scottish geologist, James Hutton, in the late 1700s and further developed by Charles Lyell. Lyell's writings had a considerable influence on Charles Darwin and his theory of evolution.

Modern geology is considered to be principally uniformitarian. Terrestrial evolution is explained in terms of plate tectonics wherein the Earth's crust is composed of a series of plates that are in almost constant motion. The collision of plates gives rise to earthquakes, volcanoes, ocean ridges and trenches, and continental drift.

However, the proposal that the extinction of the dinosaurs some 65 million years ago may be the result of a global atmospheric disaster triggered by the impact of an asteroid or comet in the Yucatan, Mexico, has contributed to the partial revival of the Catastrophist theory.

Creationism and Intelligent Design

From another perspective, we have Creationism, the belief that the universe was created by God exactly as divulged in the Book of Genesis. Creationists assert that the theory of evolution does not satisfactorily account for the existence of life on Earth and that the evidence for a suddenly created and fixed universe, rather than an evolved and changing universe, is more compelling. Most of the religious creationists are fundamentalist Christians. They believe that the Bible is the literal word of God, that the world was created in six days, and that it is just over 6,000 years old.

In scientific creationism, scientific evidence for evolution is refuted and is replaced by alternative explanations. Its followers assert that life could not have emerged from non-living matter and that the various complex species could never have developed from primitive organisms through the process of natural selection as proposed by the theory of evolution. They believe that geological features were formed by catastrophic changes and not by uniformitarian processes. There are also those in the scientific community who, while rejecting the biblical version of creation, still believe that the complexity of the cosmos implies design and therefore a designer.

Intelligent design (ID) proponents, supported mainly by religious activists and a relatively small group of non-mainstream scientists, hold the view that some forms of life are so complicated that they are best explained by the existence of an intelligent designer. An intelligent designer must have intervened to have created features such as eyes and ears and the 'irreducibly complex' biochemical processes of life.

Religious groups have been pushing for ID to be added to the science curriculum in school districts across the United States.

In November 2005, the state board of education in Kansas voted six to four to adopt science standards that involve casting doubt on evolution. However, voters in Dover, Pennsylvania removed eight members of the school board, which had been sued by parents for adding ID to the curriculum. In a ruling in December 2005, US District Judge John E Jones III found board members 'consciously chose to change Dover's biology curriculum to advance religion', violating the principle of separation of church and state. He declared that intelligent design was 'an interesting theological argument but … not science.'

The following chart shows some areas where fundamentalism and science conflict:

Where Religious Fundamentalism and Science Conflict

	Fundamentalists say	Scientists say
Teaching evolution	Schools should teach intelligent design as an alternative theory to evolution.	Evolution is the only valid explanation for the diversity of life.
Embryo research	Human life is sacred and begins at conception, so it is wrong in principle to experiment on embryos, whatever the potential pay-offs.	Research on early embryos – microscopic balls of cells – causes no pain or suffering and promises enormous medical benefits.
Condoms to prevent AIDS	Condoms promote promiscuity; the only way to stop HIV spreading is through sexual fidelity.	Evidence shows that condoms can save lives through HIV prevention.
Homosexuality	Given enough willpower and encouragement, any homosexual can renounce his or her sinful and unnatural lifestyle.	Homosexuality is not a matter of individual choice but a natural biological phenomenon.

The nineteenth-century battle between theologians and scientists regarding the opening pages of Genesis has certainly generated a new formulation of its meaning. However, it must be acknowledged that in the original mystic vision we see the first account of the universe created and unfolded by an all-empowering force. Once and for all it removed the cancer of idolatry and gave order and meaning to nature and natural phenomena. It was through progressively refined processes of observation and interpretation that we now have a more comprehensive and meaningful understanding of the vision portrayed in Genesis.

Answers to frequently asked questions from evolutionists can be found in Appendix A and from creationists in Appendix B.

Magnetism and Earth's Magnetic Field

Other natural phenomena, such as magnetism, have intrigued mankind for centuries. The earliest magnets were pieces of iron ore which possessed the property of attracting small pieces of iron and, when freely suspended, pointed approximately north and south. The direction a compass points to is the magnetic north and south and not the true north and south poles. For modern navigational or surveying purposes, the magnetic declination denotes the angular difference between the direction of true north and the direction of the Earth's field as shown by a compass. Compasses were used by the early navigators, who followed their direction without ever knowing that the Earth itself had a magnetic field.

It was not until 1600 that William Gilbert suggested the hypothesis that the Earth itself was a uniformly magnetised sphere with poles approximating to the geographic poles.

Current thinking indicates that the Earth's magnetic field originates in the fluid outer core in a dynamo-like process. The Earth's core is made up of a highly conductive iron-nickel alloy.

The Earth's magnetic field has left its mark on all rocks containing magnetic minerals. In the formation of igneous rocks (high temperature) the ferromagnetic minerals are initially nonmagnetic but become magnetic when they cool down below a certain temperature called the Curie temperature, Tc, which ranges from 200 °C to 680 °C. This type of magnetisation is called 'remanent magnetisation' and is relatively permanent.

Secondary components of magnetisation may be superimposed depending upon the history of the rock formation. These include changes in magnetic pole position resulting from continental drift in addition to reversals in the Earth's magnetic field. The last reversal was determined to have occurred approximately 780,000 years ago.

In order to determine the direction of remanent magnetisation, samples collected in the field are oriented using a magnetic compass prior to removal from the outcrop. In the laboratory, each sample is marked to preserve the initial orientation. Determination of the natural remanent magnetisation requires measurement of both moment and direction. A spinner or superconducting type of equipment may be used for this purpose. Secondary magnetisation can be removed using cleansing techniques in the laboratory, including alternating field, thermal or chemical methods. The cleansed direction of magnetisation reflects the direction of the Earth's magnetic field at the time of formation of the rock.

Practical Application

In 1963 I carried out pioneering research in the practical application of these techniques in interpreting structure.* The study was carried out on a large elliptical basin-like structure, known as the Sudbury Basin, located in Ontario, Canada.

It is interesting to note that the mechanism believed to have triggered igneous activity in this originally circular structure may have been the impact of a meteorite some 1.7 billion years ago.

Oriented samples were collected from locations all around the basin, mainly from the norite or more basic fraction that has a higher content of ferromagnetic minerals. After carrying out the

* Sopher, S, 'Palaeomagnetic Study of the Sudbury Irruptive', *Bulletin 90*, Ottawa, Geological Survey of Canada, 1963.

cleansing operation to remove 1.7 billion years of secondary magnetisation, the original remanent magnetisation was determined.

From analysing the direction of magnetisation in the north and south ranges, it was determined that, relative to the north range, the south range was folded, on average, approximately 30° northward. The east range was found to have suffered similar deformation.

The above study demonstrates just one of the methods that can be used to decipher events occurring in the remote past of the Earth's history.

Migration – Birds

The Earth's magnetic field affects many things we might not expect to be related to it. Let us now consider the fascinating field of bird migration.

The Arctic tern travels 10,000 miles (16,000 km) twice a year. The ruby-throated hummingbird travels from Canada to Central America, a distance of 1,900 miles (3,000 km). The trip may involve a non-stop flight of about 530 miles (850 km) across the Gulf of Mexico. Several other birds including the common quail, the common crane and the osprey are regular migrants.

People have noticed these migratory habits for centuries and have speculated on how these birds were able to travel such long distances to remote locations and return with such accuracy.

It appears that no single process or mechanism can account for all of the behaviours we see.

A distinction should be made between orientation and navigation. Orientation is the birds' ability to use an internal compass to accurately align themselves in an appropriate direction when released into unfamiliar surroundings. Navigation is usually thought to require some form of internal map plus the ability to orient.

Various possible orienting mechanisms were examined and experimentally tested by scientists. They included the sun compass, the star compass and the magnetic compass. However, the most successful experiments relating to bird migration have focused on the birds' ability to make use of the Earth's magnetic field. The fundamental experiments in this area were carried out by scientists in Germany in the late 1960s and 1970s. An investigation by Wolfgang and Roswitha Wiltschko* demonstrated that birds can orient to the Earth's magnetic field. A very intriguing observation in this area was the discovery of a tiny magnetic 'crystal' in the head of pigeons located between the skull and brain. It can be considered a possible biological feature that could give birds a kind of 'sixth sense'.

Migration – Sea Creatures

The migration of sea creatures is also phenomenal. The grey whale is known to travel farther than any other mammal, a distance of 12,450 miles (20,000 km) each year.

Tortoises and turtles are among the oldest living reptiles, and first appeared about 200 million years ago. The green turtle is known to travel, as a round trip, a distance of 2,800 miles (4,500 km).

Like birds, sea creatures are also believed to function under a guidance system. They appear to follow a 'magnetic map' by detecting anomalies in the Earth's magnetic field.

Other Effects of the Geomagnetic Field

The geomagnetic field may indeed affect us in many ways. From ancient wisdom to present-day superstition, the belief that the full moon holds a power over people remains strong. Scientific research shows some correlation between lunar phases and certain aspects of human behaviour, including telepathy. An experiment in the mid-1960s by neurologist Andrija Puharich† indicated that telepathy peaked during the full-moon phase, was high during the non-moon phase, but was very low during the half-moon phases.

---

* Wiltschko, *Magnetic Orientation in Animals*, New York, Springer-Verlag, 1995.
† Puharich, *Beyond Telepathy*, London, Pan Books, 1975.

While at the time it was theorised that the effects were due to the sun and moon's changing gravitational forces, as evidenced by the tides, other scientists were exploring the geomagnetic effects brought about by the sun and moon's cycles. They found that the geomagnetic forces also fluctuate during the lunar phases and are at their lowest during the full moon. This has prompted the suggestion that humans may have traces of magnetic compounds in their brains that operate in an analogous manner to that of homing pigeons. Using a technique called transcranial magnetic stimulation, magnetic fields can be used to disrupt brain functions and temporarily mimic conditions found in people with fronto-temporal dementia, a degenerative brain disease. In targeting other areas of the brain, the means may be found to actually enhance conceptual abilities. The Centre for the Mind at the University of Sydney is currently carrying out research in this field.

Survival and Adaptability

In the evolutionary history of life on Earth, the fossil record provides not only evidence of species that lived millions of years ago and have become extinct, but also those species that have survived to this day. It takes many defining characteristics to be a survivor.

A prime example is the common cockroach (suborder Blattodea), which is among the oldest winged insects living today. It first appeared in the Mississippian period over 350 million years ago; 3,500 species are known worldwide. At one time in their history they reached 6 in. in length, while a fossil uncovered recently and dating back 300 million years was approximately 4 in. long. These dark-brown to reddish-brown leathery insects have flattened oval bodies that enable them to squeeze easily through tight spaces in search of food or to escape predators. They are sensitive to vibrations and can quickly flee from danger. The head is protected by an extended shield-like pronotum (the front part of the dorsal plate of the thorax). Although they generally have two pairs of wings, they almost never use them. Nevertheless, when they are cornered, as a last resort they have been known to propel themselves at high speeds to stun the adversary. They forage for food under cover of night. Female cockroaches lay packets of twelve to twenty-five eggs in inaccessible places. Metamorphosis is simple: nymphs resemble adults. Efforts to destroy them have made some species virtually immune to pesticides. They are known to live for months on little more than dust. They are tropical insects but flourish in any environment where there is sufficient food and warmth. Contrary to popular belief, the cockroach does not spread disease.

The survival and adaptability of *la cucaracha* appears to be dependent not only on its general biological make-up, but on its unique ability to react rapidly and effectively to external stimuli. Brain size, physical prowess and aggressive behaviour do not appear to be requirements for a species' survival.

V
Science and Religion

Biological Bases

The search for the biological bases of religious, spiritual and mystical experiences has begun to bear fruit. Recent research in neurotheology, using brain scans, has targeted the neurological structures of the brain to pinpoint the areas involved in spiritual experiences. The middle temporal lobe is associated with the emotional aspects of religious experience. The frontal lobe becomes active during meditation. The lower temporal lobe facilitates prayer and meditation through images. The parietal lobes are able to give a person the feeling of oneness with the universe. It appears that the regions of the brain can cause spiritual or even mystical experiences through meditation, prayer and other stimuli such as hypnotism. Nevertheless, it would be premature at this stage to state conclusively that brain activity alone, without the effect of other factors, forms the foundation of belief.

Just as most biological functions, in advanced as well as in less developed species, are performed involuntarily, it can be convincingly demonstrated that the sense of righteousness and moral code of the great religions were in existence long before formal religious beliefs were evolved. The natural instincts of love and caring for one's own, whether carried out consciously or unconsciously, have for their goal the preservation of life – in units first, then progressively as family, as tribe and finally as a nation or even as an alliance of nations.

Evolution of Morality

The evolution of morality can be traced in the Bible. In earlier references, severe punishment and plagues were commonplace for the enemies of Israel. Burnt offerings at feast days were the custom. In later times, a more spiritual covenant, written in the heart, governed their behaviour. Sacrifices were discontinued.

Greek Philosophers and Scientists

Greek philosophers and thinkers began asking questions about what the ultimate natural units of material are made of as far back as the sixth century BCE. The concept of atomism – that the universe is composed of simple indestructible particles called atoms – was first advanced in the fifth century BCE by Leucippus and elaborated by Democratus. Later, Epicurus restated the doctrine, giving the atoms weight. Atomism was almost forgotten for over 2,200 years and was only revived in the seventeenth century. With the advancement of science, a modern version of atomism has evolved. The discovery of electrons, charged nuclei, radiation and other phenomena has obviously modified the original concept of the building blocks of matter.

Other classical Greek philosophers and scientists, dating back to Hippocrates (*c.*460–375 BCE), the father of medicine, and Aristotle (384–322 BCE), who started the science of animal biology, and Archimedes (287–212 BCE), a physicist and mathematician, based their interpretations and understanding of nature on an analysis of their observations. They discarded superstitions and the supernatural.

However, with the dominance of Christianity in the Western world, the Greek vision became dormant for a protracted period. Nevertheless, it should be noted that, at the same time, Islamic civilisations were flourishing and are credited for keeping Greek science and philosophy alive.

However, it was not until the fifteenth and sixteenth centuries, almost two millennia after Aristotle, that a veritable revival took place in science. We have an array of notables including the physician, clergyman and astronomer Galileo Galilei (1564–1642), the philosopher and mathematician Sir Isaac Newton (1642–1727), and many others. But it was the invention of both the microscope and telescope that provided the necessary tools to probe the invisible and distant worlds.

Scientific knowledge continued to expand in the seventeenth and eighteenth centuries but increased dramatically in the nineteenth and twentieth centuries and into the dawn of the twenty-first century. We believe today that nature is now knowable and predictable, and ignores the so-called 'supernatural forces' and the whim of a deity.

The Universe in Scientific Terms

Now let us take the testimony of scientists in their search and understanding of the invisible world. Although scientists do not have one model to describe everything in the universe, they have instead two basic partial theories that are logically incompatible yet give minutely accurate results: general relativity and quantum physics. General relativity accounts for gravity, the force that acts across large scales. Quantum physics describes forces that act over small scales: electromagnetism, weak forces as seen in radioactive decay, and those strong forces holding subatomic particles together.

A compelling catalyst for understanding the connection between these theories has been called the superstring theory, where elementary point particles are replaced by tiny oscillating loops called strings. Physicists have developed five different versions of this superstring theory and have synthesised them into what is called M-theory, which operates in eleven dimensions (ten space and one time). A full understanding of this mysterious M-theory, which will allow for the prediction of the behaviour of all matter and energy in all situations, is the ultimate goal.

Nevertheless, we know that matter is wholly convertible to energy and vice versa. The human body contains billions of cells and, considering the probable number of particles forming a cell, the total number of particles would be in the trillions. Where did these particles come from and where will they go?

Astronomers attempt to trace the flow of matter and energy from one form to another. A possible original source was the big bang that occurred some 13–14 billion years ago. It appears that massive stars made, essentially, most of the elements of our world. Through complex events such as supernova explosions and catastrophic collapses into black holes, accelerating cosmic rays, and other phenomena, we find that atoms are able to combine into molecules in obscure venues. Understanding the relationship of these complex events as it relates to our solar system is a prerequisite of understanding the origin of life on Earth.

Although there is no certainty in scientific theory, the eminent philosopher Karl Popper (1902–1994) stated, 'Science is perhaps the only human activity in which errors are systematically criticised and… in time, corrected.'* As an example, the Gravity Probe B, built by NASA and Stanford University at a cost of $450 million, blasted off on 6 December 2003, to test Einstein's theory of relativity, using gyroscopes to measure how the Earth's presence warps space and time.

However, as a wake-up call, scientists should heed Vine Deloria's scholarly critique of the Western belief system in his book *Evolution, Creationism and Other Modern Myths*. As the major accomplishments in physics were the result of 'inspiration' or 'intuitive mystical insight', and later translated into mathematical formulas, there is always a possibility that subjectivity rather than objectivity could play a role in the interpretation of data to fit a concept or theory.

When a human dies, and for that matter when any living organism dies, the basic particles eventually find their way to other venues and functions. It can be said that your physical body, or rather its components, may very well be recycled into either inanimate or animate repositories. A so-called physical reincarnation is in the realm of possibility.

We can relate to the voice in Longfellow's innermost heart:†

> Tell me not, in mournful numbers,
> Life is but an empty dream! –
> For the soul is dead that slumbers,
> And things are not what they seem.
>
> Life is real! Life is earnest!
> And the grave is not its goal;
> Dust thou art, to dust returnest,
> Was not spoken of the soul.

---

* Popper, *The Logic of Scientific Discovery*, London, Routledge, 1992.

† Longfellow, 'A Psalm of Life. What the Heart of a Young Man said to the Psalmist', *Knickerbocker Magazine*, October 1838.

The Soul: Religious and non-conventional Concepts

Coming to the soul of the matter, most religious beliefs are concerned with the afterlife. Both Hinduism and Buddhism share the concept of reincarnation. However, the Hindus presuppose that there is a soul, which departs from the dying body and is reborn in another body. On the other hand, Buddhists reject belief in a soul. In conventional beliefs of Chinese origin, the spirits of dead ancestors play an important role. The human soul is conceived of having two parts – the *p'o* and the *hun*. One is earthly by nature and the other heavenly by nature. In the monotheistic religions of Judaism, Christianity and Islam, the soul and its accountability in the afterlife are doctrinal.

When one contemplates the conventional soul of the major religions, there are many aspects that need clarification. For one, do all forms of life have a soul? If not, do only human beings have a soul? If so, when in time did they acquire a soul? Was it 5,000, 25,000 or 250,000 years ago, or was it from the time of Lucy or *Kenyanthropus platyops*, our possible direct ancestors of about 3.5 million years ago, or *kaddaba* of over 5.2 million years ago? Nevertheless, human beings are the only creatures able to contemplate the soul.

With regard to the mysteries of animal spirituality, Kowalski reflects:

> Are human beings the only animals that have a moral sense or conscience? Do other creatures have notions of right and wrong? We are certainly not the only beings motivated by feelings of compassion and concern for others. Altruism is widespread in the animal world.*

Animal lovers are convinced that 'man's best friend' appears to have acquired human characteristics such as loyalty and something closely resembling a conscience.

Certain species of birds such as macaws, jackdaws and wild

* Kowalski, *The Souls of Animals*, New Hampshire, Stillpoint Publishers, 1991.

geese exhibit their devotion to each other by mating for life. They also show signs of grief when they lose a mate. In the case of macaws, should one of the couple have a fatal accident, the other may commit suicide by soaring up to a great height and then closing its wings to plummet to earth.

Now, the soul or mind or spirit may not necessarily be an independent substance. However, whatever the perception, it functions through so-called matter, the human body; it is, as it were, an attribute or quality of matter. As such, its components, whether conventional or unique, or a combination of both, would proceed, as with the body, to different levels of particle and energy distribution.

I would also suggest that souls can be described as masses of energy that can assume incalculable forms and transmigrations with the capacity to communicate with each other. Assuming these perceptions are valid, are these entities currently representing their most recent departed life forms or any one of their transmigrational repositories? The latter case could be very disturbing indeed for those with the expectation of communicating exclusively with their beloved next of kin.

Television has contributed to the current phenomenal appeal of psychics, seers and mediums, who entertain viewers with their casual communication with those who have passed on.

Consequences arising from the belief in divine judgement in the afterlife must also be adequately addressed. Are sentences served for crimes committed on Earth sufficient to provide a wash before Judgement Day? What is the consequence in cases where capital punishment has been enforced, and the person executed is later found to be innocent? What should be the fate of the condemning judge and/or jury? As can be envisaged, the sequence in this line of questioning is unending and could lead to ambiguous solutions. Nevertheless, our evaluation of the substance of the values represented by the Ten Commandments in our planetary environment, and its relevance to the universe as a whole, is of paramount importance.

No matter what impact the answers to these questions might pose to theologians, the components of body as well as of soul have an eternal continuity. We are all, individually and communally, part of the evolving universe.

Survival

Are we to believe that we humans are nature's last word? We know very well from the fossil record what species went before us and when they perished, and can agree with Tennyson's dirge:

> From scarped cliff and quarried stone
> She cries, 'A thousand types are gone;
> All shall go.'*

Nevertheless, no matter to what extent the gurus, mystics and prophets differed in their views, it becomes evident that individual survival is subject to the observation of certain rules of association and behaviour. Throughout recorded history, useful customs of the past, with some modifications, have been codified into laws and rules of conduct. Most religions have endorsed and incorporated these ethical values into their fabric. The Ten Commandments of Moses and the Five Precepts of Virtues in Buddhism exemplify this. The various scriptures have also clearly included their individual sacred narratives, doctrines, rituals and traditions. Adaptations to and the bolstering of specific cultural heritages have taken place. If certain unfounded deep prejudices and intolerances can be overcome, the organisational benefits of these communal associations, coupled with their emotional involvement, have the ability and structure to stir the imagination or even to transform the lives of believers and non-believers alike.

For the last 300 years the accelerating pace of scientific advances in the fields of astronomy, cosmogony, biology, physics and even psychology had a profound effect on the foundations of established religious teaching.

The Roman Catholic Church expressed its concern in the Vatican Council of 1869–1870 and in its pronouncement against modernism in 1907. Meanwhile, the Protestant Churches of North America established themselves in the 'fundamentalism' of the Bible Belt. Similarly, in the Islamic world, militant movements such as Wahhabism and Mahdism, as well as recent fanatic

---

* Tennyson, 'In Memoriam A H H', 1833.

fundamentalist organisations, were prescribed. All these movements can be considered as symptoms not of strength, but of weakness and discredit the cause.

Ancient Civilisations – Egypt and Babylonia

However, at this time I deem it prudent to review the legacy left by the ancient civilisations and culture of Egypt and Mesopotamia in the ancient Near East. Their cultural accomplishments and beliefs strongly influenced the early Israelite tribes. Judaism and the blossoming of Christianity and Islam were the direct result. Additional philosophical issues will be addressed.

The dawn of civilisation in Egypt dates back to at least 6,500 years ago. We can readily understand why Egypt was referred to by the ancients as the 'gift of the Nile'. In this land of almost no precipitation, each and every year the Nile inundates the valley and leaves a stretch of rich alluvium on which a year's supply of food can be grown in a matter of a few months of cultivation. At that time, this phenomenon of annual flooding presented a challenge to the inhabitants. Their ingenuity led to the building of canals to channel the flood waters away from their dwellings in order to allow them to establish communities along the river. These villages eventually developed into towns and cities and centres of culture. The Nile also provided a very convenient means of rapid transport and commerce, with access to the Mediterranean, Red Sea and the Indian Ocean.

According to tradition, the deities, starting with Ra, the sun god, initially governed the Earth and influenced the development of mankind. Their task included the organisation of the physical world, the education of the people and the provision of a religious order. When the task was duly accomplished, these gods returned to Heaven and were succeeded on the throne by mortal man.

In summary, Egyptian society consisted of three estates: the monarch, the church and the people. There was in theory no limit to the power of the sovereign, or pharaoh. As the all-powerful son and successor of the gods, the centre of all government, and as

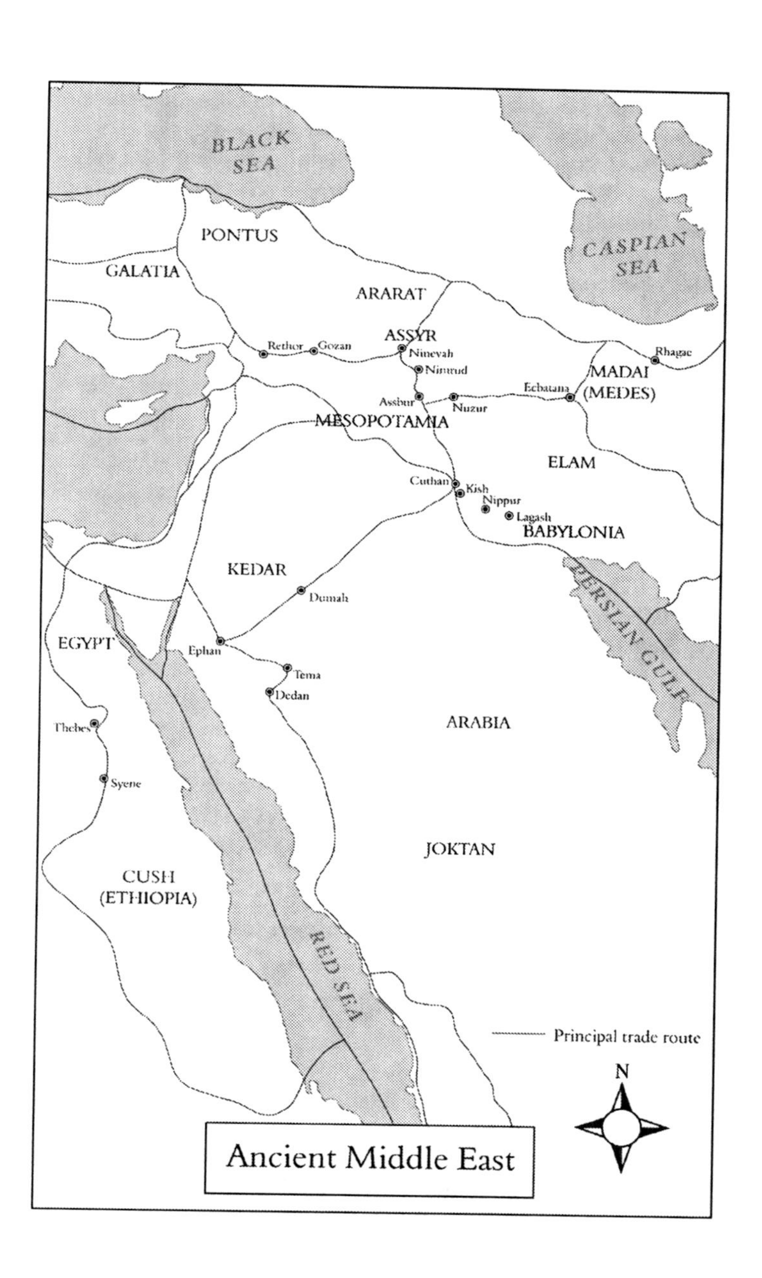

BLACK SEA
CASPIAN SEA
PONTUS
GALATIA
ARARAT
ASSYR
Rethor
Gozan
Ninevah
Nimrud
Rhagae
MADAI (MEDES)
Echatana
Asshur
Nuzur
MESOPOTAMIA
ELAM
Cuthan
Kish
Nippur
Lagash
BABYLONIA
KEDAR
Dumah
PERSIAN GULF
EGYPT
Ephan
Tema
Dedan
ARABIA
Thebes
Syene
JOKTAN
CUSH (ETHIOPIA)
RED SEA
Principal trade route
N
Ancient Middle East

personifying the unity of the life of the nation, he was not only likened to a god and called son of the gods, but actually worshipped as such. Man as far as his body was concerned, god in virtue of his soul and its properties, he was in a real and practical sense recognised as the only mediator between god and man. The victory of right over wrong was strongly emphasised. Truth and charity were highly esteemed. On one inscription is found:

> Doing that which is right, and hating that which is wrong; I was bread to the hungry, water to the thirsty, clothes to the naked, a refuge for him that was in want; that which I did to others, the great God hath done to me.*

Religious customs included the use of magic formulas and incantations to dispel evil. Circumcision was widespread, especially among the priests and higher classes.

Besides Ra, the sun god, ancient Egyptian religion deified the moon, the stars, the sacred river, trees, animals and many other objects as well as philosophical ideas. In addition, each tribe seems to have had a separate divinity – tribal monotheism. Nevertheless, at Heliopolis (City of the Sun), the chief seat of religious learning, the high priests attempted to arrange all the gods in a hierarchy with a 'Supreme Unknowable Power' at their head. According to the priestly scribes, the commoners worshipped innumerable gods or agents, but the truly initiated worshipped only the 'Power' itself. The sacred texts and inscriptions known only to the priests, taught that there was only a single Supreme Being:

> the producer of all things both in Heaven and Earth, himself not produced of any, the only true living God self-originated, who exists from the beginning.†

It was considered unlawful to represent this Being in any material form or even utter or write his name. How reminiscent of so many later biblical pronouncements! Nevertheless, this sublime contemplation did not in fact lead to what may be referred to as

* Edward Isaac Ezra, 'Civilization: Its Dawn in Egypt', paper delivered before the Union Church Literary and Social Guild, Shanghai, 1902.
† Ibid.

genuine monotheism but, more precisely, henotheism, where the followers of a religion recognise the existence of many gods, but worship only one.

Moving towards the east, we enter into Babylonia, the 'land of the Chaldees'. It lies in the fertile valley situated between the rivers Tigris and Euphrates, where powerful kingdoms and ancient civilisations once flourished. As in Egypt, the innovative inhabitants constructed a complex network of canals and irrigation ditches, which served to create bountiful harvests of wheat, millet, barley and dates, with which the land was endowed. It is of interest to note that when man first entered the alluvial plain, the shoreline of the Gulf was close to modern Baghdad.

The history of Babylonia dates back to at last 4,000 BCE and the country can be considered the foremost racial melting pot of the ancient world. Although the Sumerians were not the first people to progressively move southward down the valley as the land became dry enough to settle, they were the first to build a rich civilisation organised around a city state. Their art, architecture, economy and political acumen were impressive. Not long after the Sumerians entered the valley, the Semitic people from the west arrived to share the riches of the rivers. Empires and concepts of economic justice then appeared, notably with Sargon and Hammurabi. Later invasions from the east by the Kassites, and from the north by Assyria, caused centuries of unwanted upheaval. A vigorous but short-lived renaissance was developed by the first Chaldean king, Nabopolasser.

The range of religious beliefs and cults corresponded to the heterogeneity of the populace. From the early period, each village worshipped its own local god. However, as the empires grew, as in the case of the First Dynasty of Babylon, Marduk, the god of the city became a reigning god of all the other gods. As in Egypt, it was the priestly class who were the custodians of learning, and in some instances also the rulers of the city state. Society was carefully regulated by law, as evidenced by Hammurabi's Code, which summarised and organised social and economic life with clarity and severe justice. The sanctity of family was protected. Laws prescribed rigorous penalties for bribery, perjury, sorcery, theft, rape, kidnapping and other criminal behaviour. Among the religious observances, it is interesting to note that the Sabbath was practiced by the ancient Babylonians. An ancient inscription describes it as follows:

> The seventh day is a fast day dedicated to Merodach and Zarpanit. A lucky day. A day of rest, a Sabbath. The Shepherd of mighty nations must not eat flesh cooked at the fire or in the smoke. His clothes he must not change. White garments he must not put on. He must not offer sacrifice. The king must not drive a chariot. He must not issue royal decrees.*

In the field of scientific endeavour, astronomy is credited to have originated in Babylonia. The year was divided into months, some of twenty-nine, and some of thirty days. By means of intercalary months they brought the lunar and solar year into harmony with one another. They are also credited with devising the zodiac system, dividing the heavens into constellations of stars. The philosophers pondered over the celestial phenomena and inferred that motion implied life and, in the case of such brilliant bodies, divine life. As the sun and moon exerted such an obvious influence on seasons and human affairs, so did the gods that presided in the celestial bodies. By contemplating the nightly heavens they reached the conclusion that the configuration of the heavenly bodies had a direct relation to earthly phenomena and events.

Development of Religions of Semitic Origin

It can be seen now how this heritage of culture and religion, developed in the ancient civilisations of Egypt and Babylonia, impacted on early Israelite history referred to in the Bible. It is significant in the narrative of Abraham that he is referred to as coming from Ur of the Chaldees in Sumeria. The highest god of the Canaanites, El, is also referred to as his god.

Although the historic evolution of the beliefs of the Israelites

* Edward Isaac Ezra, 'Civilization in Chaldea', paper delivered before the Union Church Literary and Social Guild, reprinted by the *Shanghai Mercury*, Shanghai, 1903.

is not clearly defined, it is after the twelve tribes united and founded a kingdom by conquering the land of Canaan that they became truly monotheistic centring on the figure of Yahweh.

As in the development of the other religions of Semitic origin, there was a strong focus on moral or ethical concerns, and especially on justice.

Besides the Decalogue, or Ten Commandments, there were many other laws covering a variety of subjects, including the observation of feast days, sacrifices, penalties for various crimes such as robbery, murder, sexual conduct, the rite of circumcision and so on. Traditionally, there were 613 laws in total. With regard to circumcision, it is interesting to note that recent scientific research in Africa, where there is an alarming increase in the incidence of AIDS, found that the tribes that practised circumcision had a significantly lower rate of infection compared to those that did not. It was found that the foreskin was the locus for harbouring bacteria and viruses.

To bestow credence to these laws and rites that were inculcated and observed from previous civilisations and cultures, it is not altogether absurd to presume that the priests, lawmakers and scribes (*sopherim*) attributed divine authorship to them. Continuing historical as well as archaeological research has reinforced the view that at least some aspects of the text of the Bible as well as the Qur'an should be taken metaphorically rather than literally.

In Judaism and the derivative religions of Christianity, as well as in Islam, the fundamental concepts of a single divine Being, creation of the world by God, revelations to mankind by Him, and immortality after death are equally shared.

However, what differentiates Christianity from both Judaism and Islam can be summarised as follows:

a) Joshua or Jesus was the Messiah.

b) He was the Son of God.

c) He was the Saviour of mankind by His death and resurrection.

d) He is the mediator between God and man.

In a recent book, *The Pagan Christ: Recovering the Lost Light*, Tom Harpur, a Canadian Anglican theology professor, states that much of the restrictive biblical literalism preached by most churches should be rejected. He maintains that the Gospel stories of Jesus Christ are based entirely on early Egyptian and other 'pagan' mythology. Harpur's thesis is based primarily on the extensive research of the American scholar, Alvin Boyd Kuhn, who concluded that the figure of Jesus Christ is drawn from the ancient Egyptian god Horus, who also was said to have been a messiah who had a virgin birth, raised people from the dead, became incarnate in the flesh and promised everlasting life.

For obvious reasons, these deep-seated beliefs and disbeliefs will be hard to reconcile. However, considering the dual concepts of the Son of God, as well as the Immaculate Conception, a scientific definition of divine cloning may now be appropriate.

Brazil: An Example of Blending Cultures

In Brazil, the world's largest Catholic country, we have a fascinating example of how two totally different cultures have found common ground.

As a result of slavery starting in the sixteenth century, many African slaves were brought to Brazil through the ports of Salvador and Rio de Janeiro. They brought with them their African customs and religious practices such as candomblé and Umbanda. In their beliefs, there are over fifty Orixás or gods which are identified with Catholic saints. For example, Oxumare is worshipped as St Anthony, and Ogum as St George.

There is a strong overlap in membership in the two religions, and in many instances Catholic worship takes place on Sundays and candomblé and Umbanda on Mondays in the same place of worship. In the Hall of Miracles in the Church of Bonfim in Salvador, a collection of pictures and replicas of body parts of the faithful that were cured hang from the ceiling of the church. This is one form of worship where the shamans are mainly women.

It is interesting to note that although Brazil was the last country in the Americas to abolish slavery in 1888, many Afro-Brazilians returned to West Africa where they introduced their 'vibrant brand' of Catholicism, similar to that practised in Brazil. Even a replica of a church in Salvador was constructed in the small town of Agoue on the coast of Benin.

Spiritualism, Metaphysics, the Supernatural and Paranormal

At this time, awareness of other essential branches of human experience, including the complex realms of spiritualism, metaphysics and the supernatural must be appreciated.

Spiritualism, in a strict sense, can be defined as a system of thought or belief in which the possibility is maintained of mutual communication between living persons and spirits, especially those of the dead. In a broader sense, the term is applied in philosophy to any non-materialistic system.

Religions of all ages and origin contain spiritualist elements, in so far as they recognise prayer, inspiration, and the power of good or evil spirits, oracles, visions and other phenomena. It is often difficult to distinguish between paranormal and religious phenomena.

In exploring this field of largely unexplained natural phenomena, biologist Rupert Sheldrake's book – *Seven Experiments That Could Change the World: A Do-It-Yourself Guide to Revolutionary Science** – encourages the collective participation of professional scientists and non-scientists alike.

The phenomena for investigation include:

- psychic pets;
- homing pigeons;
- the organisation of termites;
- staring experiments;
- phantom limbs;
- fundamental constants; and
- experimenter expectations.

---

* Maine, Park Street Press, 2002.

Sheldrake advocates a revolutionary approach to investigate nature's hidden forces.† In the spiritual teachings of Buddhism, there is a conviction of an eternal Buddhahood resident in all beings. Through the spiritual exercises of tantra one can identify with this Buddhahood and become one with nature and the universe.

From the dawn of recorded history, 'divination' controlled the public and private life of the Sumerians and Babylonians, which in turn influenced and affected the Israelites as well as Ancient Greece and Rome. Similar practices are found in early China and India, and close parallels are recorded in ancient Mexico and Peru. Divination makes use of dreams; autohypnotism, including trance and frenzy; subconscious impression, including clairvoyance and divining rods; and necromancy, or appealing to the spirits of the dead for revealing the future or influencing the course of events.

In a broader sense, it embraces deductions from observed objects or facts. This includes palmistry, inspection of dead bodies and entrails, astrology, magical mechanisms such as suspended rings or keys, and the casting of lots.

Most of these beliefs or methods in some form or other survived and are in practice today.

Hallucinations, or seeing things that are not there, have affected the attitude of people differently in diverse cultures. In the West, from the medical standpoint, hallucinations occur when the metabolism of the brain is altered, and can be a symptom of various diseases or conditions. However, in some cultural environments, people who hallucinate are revered as spiritual leaders with ties to the spirit world.

Although it has been recognised for centuries, by psychics and mystics, that thoughts and ideas may be transmitted by telepathy, neurological experiments have only recently confirmed that these transmissions actually can and do occur between people separated by extraordinary distances.

---

† Aspects of telepathy as well as bird and sea creature migration, which may be related to Sheldrake's ideas, are considered in part IV of this work.

Parallels in Science

A parallel is found in quantum theory, which states that two particles that were once connected are always afterwards interconnected even if they become widely separated. British physicist John Bell (1928–1990) developed the mathematical criteria required to show that such a connection existed, as well as the mechanism for measuring the continued interconnection. In 1980 a French team led by physicist Alain Aspect confirmed that when two quantum particles are correlated and fired in opposite directions at the speed of light, a connection between these widely separated particles is still detected.

In discussing religion and spirituality, I must call attention to the visionary Ken Wilber, who distinguishes between two important functions of religion.* The two categories are called 'translation' and 'transformation'. In the first category we have, in essence, the creation of a sense of meaning for the self. It is this second authentic transformative spirituality that is extremely rare and precious to humanity. It is, to quote Wilber:

> Where my Master is my Self, and that Self is the Kosmos at large, and the Kosmos is my Soul – you will walk very gently into the fog of this world and transform it entirely by doing nothing at all.

Unifying Theory

Initial attempts have been made to explore the extent to which scientific parameters can correlate and embrace the different dimensions of religion. Some degree of legitimacy to this model is prompted by the solar neutrino saga. From observation at the Sudbury Neutrino Observatory in Canada, it appears that the sun's nuclear fusion is producing the required neutrinos, but that 50% of them are changing identity on their way to Earth. In order for neutrinos to be able to switch identities, some mass, however small, is required, which is contrary to the standard model of

* Wilber, *What is Enlightenment?* magazine, special tenth anniversary edition, Mass., Enlighten Next, 2001, Issue 20.

physics. Continuing research by an international team of scientists at KamLAND in central Japan have confirmed that neutrinos do indeed have some mass and can oscillate or morph from one state to another. Although they are the smallest building blocks of all matter, they are considered to contribute significantly to the total mass of the universe. As these neutrinos provide the fundamental constituents of matter they are considered to have been essential in the origin of the life-creating processes. Other predicted particles such as neutralinos and axions may also fall into this category. With this intriguing prospect in mind, a convenient religio-scientific model, utilising Einstein's general theory of relativity ($E = mc^2$) to assure appropriate legitimacy, can be contrived. It is hoped that this unifying phenomenon, portrayed by 'U', serves also as symbolism from the religious perspective. The absolute unifying theory is proposed as follows:

$A = (Um)\,c^2$ (matter relationship)
$A = UE$ (energy relationship)
Where A = Absolute

U (scientific) incorporates factors solving the paradoxes of quantum physics and other cosmic phenomena, where space, time, energy, mass, motion, and gravity are all relative.

Under special conditions, where m and E approach or reach 0 or ∞, the curling in/out of other dimensions/paradigms may result.

U (religious) is formulated so that:

- articles common to all religious beliefs and conceptions are incorporated as well as authentic transformation, i.e. purpose, ethical conduct, individual resolve, suffering, death, devotion, and praise of the diving force; and
- variables may be identified and substituted where required to represent divergences, i.e. nature, spiritual practices, astrology, reincarnation.

Additional variables may be introduced to address illuminating advances in philosophy.

The energy or matter of soul, brain, body and U are all interchangeable in the formulae for the absolute. All forms of life, matter and the sublime (capital or small s as desired) are intimately related. Varying forms and levels of consciousness in nature and the universe are included within this all-embracing framework. Subject to the value of U, we are part and parcel of nature and its fundamental forces.

The religious definition of U presents a formidable but not insurmountable challenge to theological pathfinders. Dogmas, literal interpretation and infallibility will have to be addressed.

The fundamental beliefs of the philosophy of life require assumptions and the theory of matter to integrate science and religious beliefs.

The challenge was confronted over 300 years ago by the distinguished philosopher and scholar Benedict (Baruch) Spinoza (1632–1677). He accepted René Descartes' (1596–1650) scientific and logical approach but in substance disagreed with the concept of Cartesian duality, where the world is viewed as ultimately consisting of two kinds of substances, namely mind and matter. In a nutshell, Spinoza's philosophy can be described as the most perfect form of pantheism, where God is everything and everything is God. In the context of this work, the concept of U and the absolute can be harmonised in 'panabsolutism'.

The Judaeo-Christian and Muslim world views are compatible and generally identical with the assumptions and methods of classical science of Galileo, Newton, Ampère, Faraday and Maxwell, where physical objects have an objective ongoing existence. Modern scientists hold to a view of 'quantum reality' in which objects exist or come to existence through an observation or measurement. With regards to the law of cause and effect, classical scientists use force laws to specify how one object can have an effect upon another object. Modern science envisions, on the other hand, that objects can move, emit force and emit light on a random and spontaneous basis, independent of any cause.

A framework for resolving the eternal question of whether humans created God or whether God created humans is uniquely addressed. We are all integral ingredients of the absolute, and the absolute is subject to our evolving comprehension and definition of the unifying factor. It is essential that we remove the plaque of the mind that coats and subtly infects and penetrates the very roots of the thought processes. Both scientific and religious paths to understanding require inspiration, dedication and perspiration.

Although faith may provide sustenance in times of extreme sorrow and/or anxiety, we are cautioned again by Longfellow that:

> Not enjoyment, and not sorrow,
> Is our destined end or way;
> But to act, that each to-morrow
> Find us farther than to-day.*

Pope John Paul II courageously stated in July 1999 that hell was not a physical place. In his words, 'Rather than a place, hell indicates the state of those who freely and definitively separate themselves from God, the source of all life and joy.'† With regard to Heaven, he told pilgrims that it was not 'a physical place in the clouds but a living and personal relationship of union with the Holy Trinity'.

---

* Longfellow, op. cit.

† John Paul II, *General Audience*, 28 July 1999 from the Vatican website at http://www.vatican.va/holy_father/john_paul_ii/audiences/1999/documents/hf_jp-ii_aud_28071999_en.html

VI

Education, Understanding and Cooperation

Approach

In an effort to promote greater understanding and cooperation between the peoples of the world today, a very open approach to ideological and nationalistic differences must be encouraged.

From an educational perspective, comparative religious studies should be promoted, as has been started in the UK, Scandinavia, Germany and Canada. We not only have to understand the various religions but also discover the feelings of the people of other faiths. Fundamentalism and fanaticism must be understood from the perspective of the people who hold these views. It is noteworthy that at Zeitouna, an Islamic university in Tunisia, a more tolerant and modernised form of Islam is being fostered. Topics such as comparative religion and even Darwin's theory of evolution are studied.

Psychologists and educators have indicated that for any such programme to be effective, it should be introduced as early as possible, before prejudices and biases have set in.

It is not unexpected that some religious leaders would disagree with this secular approach, as they believe this avenue would lead to legitimising and raising other systems of belief to the same stature as their own. This would lead to debasing their own 'true vision' and water down the dogma of their particular faith.

Causes and Examples of Extremism and Intolerance

Much of the ills of extremism and intolerance are the result of economic woes resulting from decades of corruption, repression and moral decay in the regions involved, and serves as a justification for the backlash of the populace.

In the particular case of Indonesia, the largest Muslim state, with a population of over 220 million people, the struggle to reform and cure Indonesia's many ills and almost total chaos after Mr Suharto's resignation in 1998 is a living example.

After Mr Suharto's resignation, democracy flowered and about thirty Islamist political parties sprang up. The belief that Islam offers the best hope of rescuing the country from political chaos was echoed almost unanimously. However, there are two models with different agendas: one is to make Indonesia an Islamic state with a constitution based on Islamic law, the other is to preserve Indonesia's tradition of secular government. The latter would mean that instead of an Islamic state, values such as a love of peace as prescribed by Islam's holy book, the Qur'an, would be upheld.

In 1999, religious tension between Muslims and Christians exploded in the Maluku Islands in eastern Indonesia, where thousands were killed. In addition, after the 11 September 2001 terrorist attack, Washington's response in Afghanistan fuelled strong anti-American and anti-Western sentiment.

The growth of unemployment and a general breakdown of law and order have created a breeding ground for radicalism. It is certainly hoped that the moderate elements in Indonesia's society prevail. Another Taliban-like episode is certainly not welcome.

In many of the other South-east Asian countries, such as Malaysia and the Philippines, which are largely Catholic, radical Islamic groups have sprung up because of discontent from real or perceived problems.

In Rwanda and the Congo, it is estimated that over 2.5 million people have been killed in recent years because of tribal conflicts. Revenge for past grievances is the primary motive.

Similar conflicts have occurred in the Balkans, where the term 'ethnic cleansing' has been used to describe the massacre of people of a different faith.

In Northern Ireland, where religious segregation between Catholics and Protestants has been and still is the norm, prejudices and ignorance are perpetuated among the youth, leading to interminable hostility among fellow countrymen.

In the 1950s, the justification for violence was romanticised by Frantz Fanon (see *Frantz Fanon: A life*, by David Macey). Frantz Fanon (1925–1961) was a French West Indian psychiatrist who argued that violence was necessary to Third World peoples, not just as a means to gain freedom but, more importantly, to heal inferiority complexes created by the colonialists. In the US he was embraced and quoted by the Black Panthers and the Weathermen. His book, *The Wretched of the Earth*, was published in Arabic and went into six editions.

The persistent struggle for the exclusive occupation and control of space, whether by a political or a religious group, is a powerful phenomenon that perhaps understanding and compassion may restrain and eventually overcome.

Other Causes of Human Suffering

Deliberate violence worldwide has resulted in the loss of millions of lives. Although it may not come as a consolation, human depravity is not the most significant cause of suffering and loss of life. Numerous epidemics, especially of bubonic plague, have had severe consequences. In the fourteenth century, a devastating outbreak known as the Black Death is estimated to have killed 25% of the population of Europe. As it is thought to have originated in Asia, countless millions of others must have been affected. A severe outbreak occurred in Hong Kong in 1894, which spread through large areas of China, India, Japan, Australia, South America, the West Indies, Madagascar, Egypt and Russia, and affected millions of people.

Epidemic outbreaks of cholera, typhoid fever, typhus fever, smallpox and other infections have caused countless other deaths. In the Middle Ages, the occurrences were regarded as manifestations of divine wrath or perhaps cosmic phenomena. Even as late as 1755, when an earthquake and resulting tsunami devastated Lisbon, Portugal, with the loss of 90,000 lives, Lisbon's leading Jesuit condemned the local officials for distributing aid and food, arguing instead that retreat into prayer was the only adequate response. This was a consequence of his belief that the devastation was caused by an angry deity. These outbreaks of disease clearly indicate that nature has no favourites. Race, creed or sex is irrelevant.

'Miracles'

Through scientific research, effective vaccines and antibiotics have been developed in the fight against viruses and bacteria. Innovative procedures such as bypass surgery and organ transplants have revolutionised the medical profession.

It is only through biomedical research that we can expect to develop therapies to treat and perhaps cure such diseases as Alzheimer's, Parkinson's and the less known but devastating Rett Syndrome, a progressive neurological disorder which occurs only in females. Even from a contemporary viewpoint and without resorting to an assumed perspective of the ancients, these 'miracles' rival those of the Bible. It can be said that more such 'miracles' occur within a twenty-four-hour period today than all the miracles in the Old and New Testaments combined.

Intelligibility and Meaning

Concurrent with the agenda for understanding and reconciling ideologies, cultures and theologies, the contribution and interaction of modern scientific research is essential.

On receiving the Templeton Prize for Progress in Religion in 2001, the Rev. Arthur Peacocke, a biochemist and Anglican priest, stated the following at a news conference:

> The search for intelligibility that characterises science and the search for meaning that characterises religion are two necessary intertwined strands of the human enterprise and are not opposed. They are essential to each other, complementary yet distinct and strongly interacting – indeed just like the two helical strands of DNA itself.

There is overwhelming agreement that the houses of worship provide the venue for healthy group association and the environment for expressing and sharing the essential emotions of love, hope, compassion, understanding and the craving for immortality. The accompaniment of music and song enhances the emotional states. However, most congregations have only a very simplistic concept of God, good and evil. Religious institutions have an obligation to educate their parishioners to a much higher level of understanding.

It is gratifying to note that in January 2002 Pope John Paul II invited more than 300 religious leaders from all of the world's major faiths to come to pray for peace in the little town of St Francis. At the meeting, Cardinal François-Xavier Van Thuan said, 'We proclaim before the world that religion must never become a reason for conflict, hatred and violence, like that seen once more in our days.'

The Pope wanted this meeting to make clear that religion is not a cause of violence, but rather a force for peace. Leaders from the major Christian denominations, as well as Jews, Hindus, Muslims, Buddhists and African animists, participated.

As far as scientific research is concerned, discoveries and development continue to solve one mystery after another. However, as with religion, the general public is almost illiterate in science, seeing or hearing headlines and not much else. Continuing, and more effective, scientific education for adults is required.

Although there can be a great amount of pleasure and joy in human existence, there is, lurking in the background of our subconscious, the awareness of the transient nature of our existence. This realisation is at least partially alleviated as the majority of the populace are adherents, nominally at least, of the major religions (see the table on p.57). Although it took *Homo sapiens* some 200,000 years to extricate itself from the more blatant forms of idolatry, it is anticipated that remarkable progress will now be achieved through the integration and non-confrontation of scientific endeavour with pan-religious concerns. However, as cautioned in the now famous remarks by Alan Greenspan about market valuations, let us also not fall victim to 'irrational exuberance' leading to an infectious 'creed' rather than his cited 'greed'.

The Balanced Life

Other seemingly mundane factors such as diet and physical activity are necessary ingredients for a more fulfilling and extended journey on Earth. The dietary laws of the Old Testament were formulated and ordained to mitigate the incidence and spread of disease in the contemporary environment of the Middle East. Kosher or clean foods include all vegetables and plants and four-footed animals that chew the cud and have parted hooves. Other clean foods include fish having both fins and scales, and certain birds and or fowl. Unclean foods included blood, and animals such as the pig, horse, birds of prey, eels and shellfish. It is obvious today that trichinosis develops when swine meat that is poorly cooked is ingested. Life-threatening diseases can also be spread by eating shellfish obtained from polluted waters. In certain types of fish poisoning, such as ciguatera and scombroid, the fish smell and taste normal and even cooking, smoking or canning will not kill the bacteria.

Physical activity does not seem to play an important role in Judaism, Christianity and Islam, most probably because of the once prevalent rural and nomadic way of life. Nevertheless, the physical prowess of Goliath and Samson was greatly admired.

However, some of the eastern religions – especially in the Upanishad version of Hinduism, yoga, the path to spiritual liberation – certain exercises were developed to establish control over the body and mind. Also, the Chinese t'ai chi chu'an (or shadow-boxing) is indebted to Taoism.

Because of today's industrialisation and mankind's sedentary lifestyle, obesity is becoming one of the biggest dangers to world health. In 2000, it was estimated by the World Health Organisation that there were 300 million obese adults worldwide, and that more than 115 million people suffer from obesity-related problems.

A prerequisite for a truly balanced life or existence is a healthy body or superstructure. This awesome superstructure houses the dynamic to spark the release of an incalculable energy potential. However, the manner in which this energy is harnessed is subject to the system in which the brain or mind is continually acculturated. Knowledge and skill and recognition of a sense of interdependence and relationship, not only with our fellow man but with the eternal universe, must be attained. A requirement for unlocking the door to higher truths and true happiness is to free ourselves of the anxieties of death by developing an awareness of our harmonious transitional relationship with the rest of the universe. From experience, both historical and personal, we can infer that, through philosophy and religion intertwined with science, the necessary enthusiasm and satisfaction in life's drive for fresher fields and newer planes of existence is attainable.

Gene Research

From another aspect of human understanding or rather misunderstanding, research in genome sequencing has given us a new and impartial vision on the question of race. Teams from independent research projects – the International Genome Sequencing Project and Celera Genomics – estimated in 2001 that the total number of human genes comes to approximately 31,000 (revised to 25,000 in 2003). The parallel studies indicate that 99.9% of the genetic code in each person is shared by the rest of the population. In fact, in some instances, there are more differences *within* a race than between races. Extending our vision in all directions and mediums, both material and spiritual, will enable us to lift the veil from entrenched emotional issues.

The issue of using cells from cloned human embryos for medical research has provoked heated debates in religious, medical and political circles. However, new research by scientists at Advanced Cell Technology in Worcester, Massachusetts, have created a monkey embryo without fertilisation through 'parthenogenesis'. Parthenogenesis is the development of an egg into an embryo without sperm fertilising the ovum, and may bypass ethical objections raised against embryo research. Research in producing human stem cells by parthenogenesis is ongoing. In addition, intriguing research at the University of Minnesota has isolated a type of versatile cell from adult bone marrow that seems capable of transforming itself into most or all of the specialised cells in the body. Experimentation by researchers working independently at the University of Edinburgh and Japan's Nara Institute of Science and Technology announced on 30 May 2003 in the journal *Cell* their discovery of a 'master gene' in embryonic stem cells that is largely responsible for giving these cells their unique regenerative and therapeutic potential. The research, conducted mainly on mouse embryo cells but also on their human equivalents, is believed to be shedding light on the mysterious capacity of embryonic stem cells to become any kind of cell the body might need, a phenomenon known as 'pluripotency'.

A new approach in microbiology referred to as systems biology goes beyond the success achieved by the sequencing of the human genome project. By focusing on showing the interconnection among cellular components and analysing a living thing as a whole and not one gene or one protein at a time is its formidable goal.

Geopolitics

In the geopolitical arena, it is essential that courses of action taken must be based on a careful appreciation of the consequences and not on the convenience of the moment. Misapplied idealism has aggravated rather than alleviated problems in the emerging countries of Asia, Africa, the Middle East and Central and South America. In the post-WWII era, ostensible aid intended for uplifting the general populace had the opposite effect in entrenching and empowering dictatorial regimes. As a current revealing example, the extremist fringe of the Salafi School of Islam, writes Roula Khalaf:

> ...believe that to destroy the primary source of oppression in the Middle East – the ruling regimes – militants must attack the American nerve centre that not only keeps the governments alive but also backs Israel, the region's 'greatest evil'.*

Similarly, at a round table discussion in February 2002, in Istanbul, foreign ministers from the European Union and the Organisation of the Islamic Conference (OIC), Hassan Hanafi, chairman of the Philosophy Department of Cairo University,

* Roula Khalaf, 'Military defeat leaves Arab resentment thriving', *Financial Times*, 19 February 2002.

speaking of 11 September, said, 'The West is responsible for fundamentalism. Modernisation equals Westernisation, equals Americanisation. There was a reaction to this.'

Repetition of any of such short-sighted stratagems may have even direr consequences in the context of the twenty-first century.

Democratisation, or the process by which individual freedom and social equality may be achieved, is crucial in tempering points of contention. The ideology of 'an eye for an eye' not only perpetuates but justifies vindictive behaviour. If biblical values are to be followed, they should be instilled by understanding and appreciating the social consequences of malevolent behaviour rather than from the standpoint of simple retribution for disobeying one of the Commandments.

Final Thoughts and Comments

As Albert Einstein so eloquently stated 'Science without religion is lame, religion without science is blind.'*

It must be acknowledged that science taken alone has its limitations. It does not satisfactorily provide a sense of coherence with regard to the meaning and purpose in one's personal life. As such, the individual quest for spirituality should not be discouraged.

Similarly, blind faith has its limitations and pitfalls. As an example, in fundamentalist Christianity, our brief sojourn on Earth is treated somewhat like a filtering or sieving process. In the protocol, only those who pass the litmus test of believing that Jesus Christ is the Son of God and Saviour of mankind, and only through Him, one will enjoy eternal life with God in Heaven. That brings to mind the question: What hellish uncertain future is in store for the remaining billions of mankind of the past and of the future? Throughout history, sectarian religions and ideologies have contributed to division, hatred and war.

Instead of attempting to find fulfilment only in the next life, appreciation of the here and now should be emphasised.

---

* Symposium of Science, Philosophy and Religion, 10 September 1941; reprinted in Einstein 1954, 41; 'Sees no Personal God', *Associated Press*, 11 September 1941.

Nevertheless, through scientific endeavour we are better able to understand the laws of nature and hopefully, through an epiphany (in its broadest sense), we will be able to appreciate their translation and/or application to the spiritual yearnings of humanity as a whole. At the same time, quoting the British scientist Richard Dawkins, the goal is to adopt 'a less human-centred view of the natural world'.*

At this stage of human development, the struggle to find evidence of meaning and a moral sense not only to cataclysmic events but to everyday experiences is of special concern and reflection. Whether death is the result of phenomena such as volcanic eruptions or deadly earthquakes and tsunamis, or the result of human atrocities, is something that has to be grappled with. At the broadest and impersonal level, it can be explained that whatever occurs in or by nature is, in essence, natural. There are no subjective moral or religious implications. Indeed, what is required is the understanding of how and why these events occur. From this perspective, the prediction and possible prevention of the painful effects of future occurrences may be realised. Nevertheless, at the personal level, humankind cannot, as yet, resign itself to this impersonal perspective. Some form of moral meaning and purpose, if only tentative, cannot be summarily dismissed. The essence of this meaning and purpose in life can be most fruitfully achieved by attempting to understand the critical aspects of our consciousness and emotions. In this respect, religions, in their various manifestations, touch a chord of recognition at a deeper level of consciousness, and by so doing bring comfort and solace to many.

However, it must be cautioned that these beliefs have not deterred a significant minority from having these same emotions stirred to the extent of condoning or even committing atrocities in the name of their faith. It is as much the responsibility of the religious leaders concerned and the public at large to rein in these anti-pluralistic and intolerant mentalities. At the present time, more then ever, reconciliation is required to temper the emotions fired by fundamentalists and hypocritical extremists.

As a footnote to the recognition of religion as a bastion of

---

* Richard Dawkins, *The Selfish Gene*, New York, Oxford University Press, 1989.

comfort and solace, it should be noted that there are those who regard belief in God and immortality as illusions. Bertrand Russell, in an essay entitled 'What I Believe', put it very succinctly:

> I believe that when I die I shall rot, and nothing of my ego will survive. I am not young, and I love life. But I should scorn to shiver with terror at the thought of annihilation. Happiness is nonetheless true happiness because it must come to an end, nor do thought and love lose their value because they are not everlasting. Even if the open windows of science at first make us shiver after the cosy indoor warmth of traditional humanising myths, in the end the fresh air brings vigour, and the great spaces have a splendour of their own.*

Now, coming back to nature and evolution, we can surmise that humans have ample room and opportunity to climb the evolutionary ladder to the next level of development and adaptation. However, of the millions of species that have inhabited our Earth and then disappeared, the average survival time of a particular species is estimated at between 5 and 10 million years. Statistically speaking, our species has not much more than 5 million years to go. In any event, to quote J T Trevors, 'There is no end goal or linear end point to evolution, it just happens.'†

It is also evident that the effects of dramatic climatic changes affecting sea level elevations and inhospitable environments can limit our adaptability. Recent studies have suggested that climate warming is linked to the faster and farther spread of pathogens. Patterns of the spread of diseases in many forms of wildlife have been documented. Mankind's voracious appetite for natural resources may also limit the long-range sustainability of modern economies.

In addition, extraterrestrial collision and other natural phenomena could make us just another experiment or offshoot in the fossil record. As a chilling reminder, an asteroid with a diameter of between 50 and 120 yards had what scientists believe

---

* Russell, 'What I Believe', *Why I am not a Christian and other essays*, London and New York, Simon and Schuster, 1967.

† Trevors, 'The Theory of Everything and Molecular Evolution', *World Journal of Microbiology and Biotechnology*, Mass., Kluwer Academic Publishers, 1998, vol. 14.

to be one of the closest known approaches to Earth by objects of this size, in June 2002. However, we can be assured that eventually, but most assuredly long before our sun becomes a 'red giant' and engulfs Earth, life in all its forms will have terminated. Nevertheless, our journey from eternal nothingness to somethingness and beyond will continue in all directions and dimensions.

Meanwhile, research by particle physicists and astronomers will one day lead to the unification of the four forces of nature (electromagnetism, weak force, strong force and gravity). At the same time, if all four forces were in fact a single force that assumed different complexions after the big bang, the forces and particles existing today had no initial meaning. In addition, in the zero-sum scenario where the negative energy of all the gravitational fields in the universe is offset by all the positive energy of all matter in the universe, '...the universe could come from nothing because it is, fundamentally nothing'.*

Nevertheless, the continual flow and conversion or morphing of the various states of energy and matter with the self-assembly and disassembly of basic elements and compounds has resulted in our very existence in this awesome universe, and as such will continue in some state to eternity.

The accelerating explosion of scientific development will certainly present a formidable but hopefully a not insurmountable challenge to future philosophers, poets and theological visionaries. Breakthroughs in both science and religion have been and will continue to be the result of inspiration or what Ian G Barbour refers to as 'creative imagination'.

As has been illustrated and insinuated, we are all inheritors of the brilliant and divine eternal past, are immersed in the fleeting present, and are on the springboard to the brilliance and divinity of eternity.

---

* Lemley, 'Guth's Grand Guess', *Discover*, New York, April 2002.

Appendix A

Frequently Asked Questions

What is evolution?

Evolution, in the broadest sense, explains that what we see today is different from what existed in the past. Galaxies, stars, the solar system, and Earth have changed through time, and so has life on Earth.

Biological evolution concerns changes in living things during the history of life on Earth. It explains that living things share common ancestors. Over time, biological processes such as natural selection give rise to new species. Darwin called this process 'descent with modification', which remains a good definition of biological evolution today.

Isn't evolution just an inference?

No one saw the evolution of one-toed horses from three-toed horses, but that does not mean that we cannot be confident that horses evolved. Science is practised in many ways besides direct observation and experimentation. Much scientific discovery is done through indirect experimentation and observation, in which inferences are made, and hypotheses generated from those inferences are tested.

For instance, particle physicists cannot directly observe sub-atomic particles because the particles are too small. They make inferences about the weight, speed, and other properties of the particles based on other observations. A logical hypothesis might be something like this: If the weight of this particle is Y, when I bombard it, X will happen. If X does not happen, then the hypothesis is disproved. Thus, we can learn about the natural world even if we cannot directly observe a phenomenon – and that is true about the past, too.

In historical sciences like astronomy, geology, evolutionary biology and archaeology, logical inferences are made and then tested against data. Sometimes the test cannot be made until new data are available but a great deal has been done to help us understand the past. For example, scorpion flies (Mecoptera) and true flies (Diptera) have enough similarities that entomologists consider them to be closely related. Scorpion flies have four wings of about the same size, and true flies have a large front pair of wings, but the back pair is replaced by small club-shaped structures. If two-winged flies evolved from ancestors like scorpion flies, as comparative anatomy suggests, then an intermediate true fly with four wings should have existed – and in 1976 fossils of such a fly were discovered. Furthermore, geneticists have found that the number of wings in flies can be changed through mutations in a single gene.

Something that happened in the past is thus not 'off-limits' for scientific study. Hypotheses can be made about such phenomena, and these hypotheses can be tested and can lead to solid conclusions Furthermore, many key mechanisms of evolution occur over relatively short periods and can be observed directly – such as the evolution of bacteria resistant to antibiotics.

Evolution is a well-supported theory drawn from a variety of sources of data, including observations about the fossil record, genetic information, the distribution of plants and animals, and the similarities across species of anatomy and development. Scientists have inferred that descent with modification offers the best scientific explanation for these observations.

Is evolution a fact or a theory?

The theory of evolution explains how life on Earth has changed. In scientific terms, 'theory' does not mean 'guess' or 'hunch', as it does in everyday usage. Scientific theories are explanations of natural phenomena built up logically from testable observations and hypotheses. Biological evolution is the best scientific explanation we have for the enormous range of observations about the living world.

Scientists most often use the word 'fact' to describe an observation. But scientists can also use 'fact' to mean something that has been tested or observed so many times that there is no longer a compelling reason to keep testing or looking for examples. The occurrence of evolution in this sense is a fact. Scientists no longer question whether descent with modification occurred because the evidence supporting the idea is so strong.

Don't many famous scientists reject evolution?

No. The scientific consensus around evolution is overwhelming. Those opposed to the teaching of evolution sometimes use quotations from prominent scientists out of context to claim that scientists do not support evolution. However, examination of the quotations reveals that the scientists are actually disputing some aspect of *how* evolution occurs, not if evolution occurred. For example, the biologist Stephen Jay Gould once wrote that 'the extreme rarity of transitional forms in the fossil record persists as the trade secret of palaeontology'. But Gould, an accomplished palaeontologist and eloquent educator about evolution, was arguing about how evolution takes place. He was discussing whether the rate of change of species is slow and gradual or whether it takes place in bursts after long periods when little change occurs – an idea known as punctuated equilibrium. As Gould writes:

> This quotation, although accurate as a partial citation, is dishonest in leaving out the following explanatory material showing my true purpose – to discuss rates of evolutionary change, not to deny the fact of evolution itself.

Gould defines 'punctuated equilibrium' as follows:

> Punctuated equilibrium is neither a creationist idea nor even a non-Darwinian evolutionary theory about sudden change that produces a new species all at once in a single generation. Punctuated equilibrium accepts the conventional idea that new species form over hundreds or thousands of generations and through an extensive series of intermediate stages. But geological time is so long that even a few thousand years may appear as a mere 'moment' relative to the several million years of existence for most species. Thus, rates of evolution vary enormously and new species may appear to arise 'suddenly' in geological time, even

though the time involved would seem long, and the change very slow, when compared to a human lifetime.

If humans evolved from apes, why are there still apes?

Humans did not evolve from modern apes, but humans and modern apes shared a common ancestor, a species that no longer exists. Because we share a recent common ancestor with chimpanzees and gorillas, we have many anatomical, genetic, biochemical, and even behavioural similarities with these African great apes. We are less similar to the Asian apes – orang-utans and gibbons – and even less similar to monkeys, because we share common ancestors with these groups in the more distant past.

Evolution is a branching or splitting process in which populations split off from one another and gradually become different. As the two groups become isolated from each other, they stop sharing genes, and eventually genetic differences increase until members of the groups can no longer interbreed. At this point, they have become a separate species. Through time, these two species might give rise to new species, and so on through millennia.

Why can't we teach creation science in my school?

The courts have rules that 'creation science' is actually a religious view. Because public schools must be religiously neutral under the US Constitution, the courts have held that it is unconstitutional to present creation science as legitimate scholarship.

In particular, in a court case in which supporters of creation science testified in support of their view, a district court declared that creation science does not meet the tenets of science as scientists use the term (*McLean v Arkansas Board of Education*). The Supreme Court has held that it is illegal to require that creation science be taught when evolution is taught (*Edwards v Aguillard*). In addition, district courts have decided that individual teachers cannot advocate creation science on their own (*Peloze v San Jan Capistrano School District* and *Webster v New Lennox School District*). (See *Teaching About Evolution and the Nature of Science*, Appendix A,

National Academy of Sciences, Washington DC, 1998.)

Teachers' organisations such as the National Science Teachers' Association, the National Association of Biology Teachers, the National Science Education Leadership Association, and many others, have rejected the science and pedagogy of creation science and have strongly discouraged its presentation in public schools. In addition, a coalition of religious and other organisations has noted in 'A Joint Statement of Current Law' that 'in science class, [schools] may present only genuinely scientific critiques of, or evidence for, any explanation of life on Earth, but not religious critiques (beliefs unverifiable by scientific methodology)'. (See *Teaching About Evolution and the Nature of Science*, Appendices B and C, National Academy of Sciences, Washington DC, 1998.)

Some argue that 'fairness' demands the teaching of creationism along with evolution. But a science curriculum should cover science, not the religious views of particular groups or individuals.

If evolution is taught in schools, shouldn't creationism be given equal time?

Some religious groups deny that micro-organisms cause disease, but the science curriculum should not therefore be altered to reflect this belief. Most people agree that students should be exposed to the best possible scholarship in each field. That scholarship is evaluated by professionals and educators in those fields. Scientists as well as educators have concluded that evolution – and only evolution – should be taught in science classes because it is *the only scientific explanation* for why the universe is the way it is today.

Many people say that they want their children to be exposed to creationism in school, but there are thousands of different ideas about creation among the world's people. Comparative religions might comprise a worthwhile field of study, but not one appropriate for a science class. Furthermore, the US Constitution states that schools must be religiously neutral, so legally a teacher cannot present any particular creationist view as being more 'true' than others.

Appendix B*

Do You Believe that Evolution is True?

If so, then provide an answer to the following questions. 'Evolution' in this context is the idea that natural, undirected processes are sufficient to account for the existence of all natural things.

Something from nothing?

The big bang, the most widely accepted theory of the beginning of the universe, states that everything developed from a small dense cloud of subatomic particles and radiation which exploded, forming hydrogen (and some helium) gas. Where did this energy/matter come from? How reasonable is it to assume it came into being from nothing? And even if it did come into being, what would cause it to explode?

We know from common experience that explosions are destructive and lead to disorder. How reasonable is it to assume that a 'big bang' explosion produced the opposite effect – increasing 'information', order and the formation of useful structures, such as stars and planets, and eventually people?

Physical laws an accident?

We know the universe is governed by several fundamental physical laws, such as electromagnetic forces, gravity, conservation of mass and energy, etc. The activities of our universe depend upon these principles like a computer program depends upon the existence of computer hardware with an instruction set. How reasonable is it to say that these great controlling principles developed by accident?

---

* Adapted from National Academy of Sciences, *Teaching About Evolution and the Nature of Science*, Washington, DC, National Academy Press, 1998.

Order from disorder?

The second law of thermodynamics may be the most verified law of science. It states that systems become more disordered over time, unless energy is supplied and directed to create order. Evolutionists say that the opposite has taken place – that order *increased* over time without any directed energy. How can this be?

Aside: Evolutionists commonly object that the second law of thermodynamics applies to closed, or isolated systems, and that the Earth is certainly not a closed system (it gets lots of raw energy from the sun, for example). However, all systems, whether open or closed, tend to deteriorate. For example, living organisms are open systems but they all decay and die. Also, the universe in total is a closed system. To say that the chaos of the big bang has transformed itself into the human brain with its 120 trillion connections is a clear violation of the second law of thermodynamics.

We should also point out that the availability of raw energy to a system is a necessary but far from sufficient condition for a local decrease in entropy to occur. Certainly the application of a blow torch to bicycle parts will not result in a bicycle being assembled; only the careful application of directed energy will, such as from the hands of a person following a plan. The presence of energy from the sun does *not* solve the evolutionist's problem of how increasing order could occur on the Earth, contrary to the second law of thermodynamics.

Information from randomness?

Information theory states that 'information' never arises out of randomness or chance events. Our human experience verifies this every day. Hew can the origin of the tremendous increase in information from simple organisms up to man be accounted for? Information is always introduced from the outside. It is impossible for natural processes to produce their own actual information or meaning, which is what evolutionists claim has happened. Random typing might produce the string 'dog', but it only means something to an intelligent observer who has applied a definition to this sequence of letters. The generation of information always requires intelligence, yet evolution claims that no intelligence was involved in the ultimate formation of a human being whose many systems contain vast amounts of information.

Life from dead chemicals?

Evolutionists claim that life formed from non-life (dead chemicals), so-called 'abiogenesis', even though it is a biological law ('biogenesis') that life only comes from life. The probability of the simplest imaginable replicating system forming by itself from non-living chemicals has been calculated to be so very small as to be essentially zero – much less than on chance in the number of electron-sized particles that could fit in the entire visible universe! Given these odds, is it reasonable to believe that life formed itself?

Complex DNA and RNA by chance?

The continued existence (the reproduction) of a cell requires DNA (the 'plan') and RNA (the 'copy mechanism'), both of which are tremendously complex. How reasonable is it to believe that these two co-dependent necessities came into existence by chance at exactly the same time?

Life is complex

We know and appreciate the tremendous amount of intelligent design and planning that went into landing a man on the moon. Yet the complexity of this task pales in comparison to the complexity of even the simplest life form. How reasonable is it to believe that purely natural processes – with no designer, no intelligence, and no plan – produced a human being?

Where are the transitional fossils?

If evolution has taken place our museums should be overflowing with the skeletons of countless transitional forms. Yet after over one hundred years of intense searching, only a small number of transitional candidates are touted as proof of evolution. If evolution has really taken place, where are the transitional forms?

And why does the fossil record actually show all species first appearing fully formed, with most nearly identical to current instances of the species?

Aside: Most of the examples touted by evolutionists concentrate on just one feature of the anatomy, like a particular bone or the skull. A true transitional fossil should be intermediate in many if not all aspects. The next time someone shows you how a bone changed over time ask them about the rest of the creature too!

Many evolutionists still like to believe in the 'scarcity' of the fossil record. Yet simple statistics will show that, given you have found a number of fossil instances of a creature, the chances that you have missed every one of its imagined predecessors is very small. Consider the trilobites for example. These fossils are so common you can buy one for under $20, yet no fossils of a predecessor have been found!

Could an intermediate even survive?

Evolution requires the transition from one kind to another to be gradual. And don't forget that 'natural selection' is supposed to retain those individuals which have developed an advantage of some sort. How could an animal intermediate between one kind and another even survive (and why would it ever be selected?), when it would not be well suited to either its old environment or its new environment? Can you even imagine a possible sequence of small changes which takes a creature from one kind to another, all the while keeping it not only alive, but improved?

Aside: Certainly a 'light-sensitive spot' is better than no vision at all. But why would such a spot even develop? (Evolutionists like to take this for granted.) And even if it did develop, to believe that mutations of such a spot eventually brought about the tremendous complexities of the human eye strains all common sense and experience.

Reproduction without reproduction?

A main tenet of evolution is the idea that things develop by an (unguided) series of small changes, caused by mutations, which are 'selected', keeping the 'better' changes over a very long period of time. How could the ability to reproduce evolve, without the ability to reproduce? Can you even imagine a theoretical scenario which would allow this to happen? And why would evolution produce two sexes, many times over? Asexual reproduction would seem to be more likely and efficient!

Aside: To relegate the question of reproduction to 'abiogenesis' does *not* address the problem. To assume existing, reproducing life for the principles of evolution to work on is a *huge* assumption which is seldom focused on in popular discussions.

Plants without photosynthesis?

The process of photosynthesis in plants is very complex. How could the first plant survive unless it already possessed this remarkable capability?

How do you explain symbiotic relationships?

There are many examples of plants and animals which have a 'symbiotic' relationship (they need each other to survive). How can evolution explain this?

It's no good unless it's complete

We know from everyday experience that an item is not generally useful until it is complete, whether it be a car, a cake or a computer program. Why would natural selection start to make an eye, an ear or a wing (or anything else) when this item would not benefit the animal until it was completed?

Aside: Note that even a 'light-sensitive spot' or the simplest version of any feature is far from a 'one-jump' change that is trivial to produce.

Explain metamorphosis!

How can evolution explain the metamorphosis of the butterfly? Once the caterpillar evolves into the 'mass of jelly' (out of which the butterfly comes), wouldn't it appear to be 'stuck'?

It should be easy to show evolution

If evolution is the grand mechanism that has produced all natural things from a simple gas, surely this mechanism must be easily seen. It should be possible to prove its existence in a matter of weeks or days, if not hours. Yet scientists have been bombarding countless generations of fruit flies with radiation for several decades in order to show evolution in action and still have only produced... more (deformed) fruit flies. How reasonable is it to believe that evolution is a fact, when even the simplest of experiments has not been able to document it?

Aside: The artificial creation of a new species is far too small a change to prove that true 'macro-evolution' is possible. A higher-order change, where the information content of the organism has been increased, should be demonstrable but is not. Developing a new species changes the existing information, but does not add new information, such as would be needed for a new organ, for example.

Complex things require intelligent design, folks!

People are intelligent. If a team of engineers were to one day design a robot which could cross all types of terrain, could dig large holes, could carry several times its weight, found its own energy sources, could make more robots like itself, and was only one-eighth of an inch tall, we would marvel at this achievement. All of our life's experiences lead us to know that such a robot could never come about by accident, or assemble itself by chance, even if all of the parts were available and lying next to each other. And we are certain beyond doubt that a canister of hydrogen gas, no matter how long we left it there or what type of raw energy we might apply to it, would never result in such a robot being produced. But we already have such a 'robot' – it is called an 'ant', and we squash them because they are 'nothing' compared to people. And God made them, and he made us. Can there be any other explanation?

Appendix C

Earth Fact Sheet

Bulk parameters

Mass (10^{24} kg)	5.9736
Volume (10^{10} km^3)	108.321
Equatorial radius (km)	6378.1
Polar radius (km)	6356.8
Volumetric mean radius (km)	6371.0
Core radius (km)	3485
Ellipticity (flattening)	0.00335
Mean density (kg/m^3)	5515
Surface gravity (m/s^2)	9.78
Escape velocity (km/s)	11.186
GM (x 106 km^3/s)	0.3986
Bond albedo	0.306
Visual geometric albedo	0.367
Visual magnitude V (1,0)	-3.86
Solar irradiance (W/m^2)	1367.6
Black-body temperature (K)	254.3
Topographic range (km)	20
Moment of inertia (I/MR^2)	0.3308
J_2 (x 10^{-6})	1082.63
Number of natural satellites	1
Planetary ring system	None

Orbital parameters

Semimajor axis (10^6 km)	149.60
Sidereal orbit period (days)	365.256
Tropical orbit period (days)	365.242
Perihelion (10^6 km)	147.09
Aphelion (10^6 km)	152.10
Mean orbital velocity (km/s)	29.78
Maximum orbital velocity (km/s)	30.29
Minimum orbital velocity (km/s)	29.29
Orbit inclination (°)	0.000
Orbit eccentricity	0.0167
Sidereal rotation period (hrs)	23.9345
Length of day (hrs)	24.0000
Obliquity to orbit (°)	23.45

Earth mean orbital elements (J2000)

Semimajor axis (AU)	1.00000011
Orbital eccentricity	0.01671022
Orbital inclination (°)	0.00005
Longitude of ascending node (°)	-11.26064
Longitude of perihelion (°)	102.94719
Mean longitude (°)	100.46435

North pole of rotation

Right ascension (T)	0.00 – 0.641*
Declination (T)	90.00 – 0.557

Terrestrial magnetosphere

Dipole field strength (gauss-Re†)	0.3076
Latitude of dipole (°)	N: 78.6
Longitude of dipole (°)	N/70.1
Dipole offset (planet centre to dipole centre) distance	0.0
Latitude of offset vector (°)	N: 18.3
Longitude of offset vector (°)	N/147.8

Terrestrial atmosphere

Surface pressure (mb)	1014
Surface density (kg/m^3)	1.217
Scale height (km)	8.5
Average temperature (°K/°C)	288/15
Diurnal temperature range (°K/°C)	283–293/10–20
Wind speeds (m/s)	0–100
Mean molecular weight (g/mole)	28.97

* T = Julian centuries from reference date. Reference date is 1.5 Jan 2000 (JD 2451545.0).
† Re denotes Earth radii 6,378 km.

Atmospheric composition (by volume, dry air)

Major	%
Nitrogen (N_2)	78.084
Oxygen (O_2)	20.946
Minor	Parts per million (ppm)
Argon (Ar)	9340
Carbon Dioxide (CO_2)	3.83
Neon (Ne)	18.18
Helium (He)	5.24
CH_4	1.75
Krypton (Kr)	1.14
Hydrogen (H_2)	0.55
Water	Highly variable typically makes up about 1%.

Bibliography

Barbour, Ian G, *Religion and Science*, seventh edition, London, Harper-Collins, 1997

Booth, Nicholas, *Exploring the Solar System*, New York, Cambridge University Press, 1996

Burke, Patrick T, *The Major Religions*, Oxford, Blackwell, 1986

Carr, Michael H [ed.], *The Geology of the Terrestrial Planets*, NASA, Washington DC, 1984

Carruth, William Herbert, Each to His Own Tongue, *The New England Magazine Co.*, 1895, vol. 19, p.323

Dawkins, Richard, *The Selfish Gene*, New York, Oxford University Press, 1989

Deloria Jr, Vine, *Evolution, Creationism and Other Modern Myths*, Golden, Colorado, Fulcrum Publishing, 2002

Engelbert, Phyllis, and Dupuis, Diane L, *The Handy Space Answer Book*, Canton, MI, Gale, 1998

Ernst, Wallace Gary, *The Dynamic Planet*, New York, Columbia University Press, 1990

Ezra, Edward Isaac, 'Civilization: Its Dawn in Egypt', paper delivered before the Union Church Literary and Social Guild, Shanghai, 1902

——, 'Civilization in Chaldea', paper delivered before the Union Church Literary and Social Guild, reprinted by the *Shanghai Mercury*, Shanghai, 1903

Frege, Gottlob, *Begriffschrift* (Concept Script), second edition, Hildescheim, G Olm, 1964 (1879)

Gray, Russell, D and Atkinson, Quentin D, 'Language-tree divergence times support the Anatolian theory of Indo-European origin', *Nature*, 27 November 2003, vol. 426, no. 6965, pp.435–438

Harpur, Tom, *The Pagan Christ: Recovering the Lost Light*, Toronto,

Thomas Allen, 2004

Henbest, Nigel, *The Planets*, first edition, London, Penguin, 1992

Khalaf, Roula, 'Military defeat leaves Arab resentment thriving', *Financial Times*, 19 February 2002

Kowalski, Gary, *The Souls of Animals*, New Hampshire, Stillpoint Publishers, 1991

Lemley, Brad, 'Guth's Grand Guess', *Discover*, New York, April 2002

Lemonick, Michael D and Dorfman, Andrea, 'The Dawn of Man', *Time*, 23 July 2001, vol. 158, no. 3

Longfellow, Henry Wadsworth, 'A Psalm of Life. What the Heart of a Young Man said to the Psalmist', *Knickerbocker*, October 1838

Magee, Bryan, *The Story of Philosophy*, London, Dorling Kindersley, 1998

Man, John, 'The Birth of Our Planet', for *Reader's Digest*, reprinted New York, 1999

Miller, Russell, *Planet Earth*, Morristown, New Jersey, Time-Life Books, 1983

NASA, *Cosmic Journeys, to the Edge of Gravity, Space, and Time*, NASA, 1999

Popper, Karl R, *The Logic of Scientific Discovery*, London, Routledge, 1992

Puharich, Andrija, *Beyond Telepathy*, London, Pan Books, 1975

Russell, Bertrand, 'What I Believe', *Why I am not a Christian and other essays*, London and New York, Simon and Schuster, 1967

Sheldrake, Rupert, *Seven Experiments That Could Change the World: A Do-it-yourself Guide to Revolutionary Science*, Maine, Park Street Press, 2002

Smart, Ninian, *The World's Religions*, second edition, London, Cambridge University Press, 1998

Sopher, David E, *Geography of Religions*, New Jersey, Prentice Hall, 1967

Sopher, Stephen R, 'Palaeomagnetic Study of the Sudbury

Irruptive', *Bulletin 90*, Ottawa, Geological Survey of Canada, 1963

Stokes, William Lee, *Essentials of Earth History*, fourth edition, New Jersey, Prentice Hall, 1982

Tennyson, Lord Alfred, 'In Memoriam A H H', 1833

Trevors, Jack, 'The Theory of Everything and Molecular Evolution', *World Journal of Microbiology and Biotechnology*, Mass., Kluwer Academic Publishers, 1998, vol. 14

Trevors, Jack and Psenner, Roland, *Federation of European Microbiological Societies Microbiology Reviews*, Elsevier Science, BV, 2001, vol. 25, pp.573–582

Van Andel, Tjeerd H, *New Views on an Old Planet*, New York, Cambridge University Press, 1985

Warren, Rick, *The Purpose Driven Life*, Grand Rapids, Michigan, Zondervan, 2002

Whitehead, Alfred North and Russell, Bertrand, *Principia Mathematica*, second edition, London, Cambridge University Press, 1935

Wilber, Ken, *What is Enlightenment?*, special tenth anniversary edition, Mass., Enlighten Next, 2001, Issue 20

Wiltschko, Wolfgang and Roswitha, *Magnetic Orientation in Animals*, New York, Springer-Verlag, 1995

Wittgenstein, Ludwig, *Philosophical Investigations*, third edition, translated by G E M Anscombe, edited by Anscombe and Rhees, New York, Macmillan, 1958

Wong, Kate et al., 'New Look at Human Evolution', *Scientific American*, special edition, 13:2, 2003

Index

Printed in the United States
109169LV00001B/99/A

9 781847 480040